Essays in Biochemistry

Essays in Biochemistry

Edited for The Biochemical Society by

R. D. Marshall

Division of Biochemistry
Department of Bioscience and Biotechnology
University of Strathclyde
Glasgow
Scotland

K. F. Tipton

Department of Biochemistry
Trinity College
University of Dublin
Dublin 2
Eire

Volume 22

1986

Published for The Biochemical Society by Academic Press
London, Orlando, San Diego, New York, Austin
Boston, Sydney, Tokyo, Toronto

ACADEMIC PRESS INC. (LONDON) LTD.
24/28 Oval Road
London NW1 7DX

U.S. Edition published by
ACADEMIC PRESS INC.
Orlando
Florida 32887

ISBN 0-12-158122-5
ISSN 0071-1365

Photoset by Paston Press, Norwich
Printed in Great Britain by
Whitstable Litho Ltd,
Whitstable, Kent

Biography

Len Hall, a Leeds graduate, obtained his PhD in the Biochemistry Department at Leicester University. After completing a two year Royal Society post-doctoral fellowship at the University of Berne, he joined Professor Roger Craig and Professor Peter Campbell's group at the Middlesex Hospital Medical School, to work on various aspects of the molecular biology of the mammary gland. In 1983 he became a 'New Blood' Lecturer in Biochemistry at the University of Bristol where his research interests are mainly centred around the hormonal control of gene expression.

Peter N. Campbell graduated in Chemistry from University College, London in 1942, and obtained his PhD in Biochemistry in 1949 at the same College. He then worked at the National Institutes for Medical Research, Mill Hill, London, with Dr T. S. Work, mainly on the subject of protein synthesis, moving to the Middlesex Hospital Medical School in 1954. In 1961 he became Chairman of the Department of Biochemistry at the University of Leeds, but he moved back to London in 1976 when he was appointed Director of the Courtauld Institute of Biochemistry. His research work has mainly concerned the biosynthesis of various proteins in animal tissues, with particular emphasis on milk proteins; but he was also instrumental in the discovery of the autoimmune basis of Hashimoto's disease. He was a founder editor of *Essays in Biochemistry*.

Kenneth B. M. Reid obtained his PhD in Biochemistry at Aberdeen University, in 1968, for studies on the structure and biosynthesis of fish proinsulin and its conversion to insulin. In 1969 he was awarded an ICI Research Fellowship to work in the Biochemistry Department at Oxford University. He joined the MRC Immunochemistry Unit in Oxford in 1971 and has continued to work there to the present date, becoming Director of the Unit in 1986. His research at Oxford has involved the structure/function relationships of the proteins of the human complement system and the molecular cloning of the DNA coding for these proteins.

Tony Turner graduated in 1969 with a triple First in Natural Sciences at the University of Cambridge and remained in the Department of Biochemistry in order to undertake postgraduate research with Keith Tipton. After a year as Royal Society European Fellow at the Mario Negri Pharmacology Research Institute in Milan, he returned in 1973 to a Lectureship in the

Department of Biochemistry, University of Leeds. He is currently Senior Lecturer and joint Group Leader of the MRC Membrane Peptidase Research Group in Leeds. He is also currently Honorary Secretary of the Neurochemical Group of the Biochemical Society and, from July 1987, he will be Chairman of the Editorial Board of the Biochemical Journal. His main research interests are in the biochemistry of neuropeptides and of neurotransmitter amino acids.

Alexander P. Demchenko graduated in molecular physics in 1967 from the Kiev State Shevchenko University. He worked at the Institute of Biochemistry of the Ukrainian Academy of Sciences under Professor V. A. Belitser on mechanisms of fibrin polymerization and gained a Candidate of Science degree in Biochemistry in 1972. His subsequent work was devoted to development of spectroscopic techniques in studies of proteins. From 1976 until 1982 he worked at the Institute of Gerontology (Kiev) on the problem of structural modifications of proteins on ageing and gained a Doctor of Science degree in Biophysics. He is currently working at the A. V. Palladin Institute of Biochemistry and is head of the Laboratory of Membrane Biophysics. His research interests include intramolecular protein dynamics and protein-membrane interactions.

Henry McIlwain's initial work was in commercial chemical laboratories and he subsequently held research awards at Newcastle on Tyne and Oxford concerning the organic chemistry of natural products and of free radicals, especially of phenazine derivatives (with G. R. Clemo and R. Robinson). He joined Medical Research Council units in 1937, first in bacterial chemistry and later in cell metabolism (with P. Fildes and H. A. Krebs), and subsequently held biochemical posts in Sheffield and with the London County Council Mental Health Services. At the Institute of Psychiatry, London, he was Professor of Biochemistry from 1955 to 1980 and initiated much neurochemical work, especially concerning the metabolism, the excitation and the functioning of neural systems. He was a founder-member of the Society for General Microbiology, of the International Society for Neurochemistry (ISN) and of the Neurochemical Group of the Biochemical Society. His present post-retirement work in neurochemistry is in the Division of Biochemistry, UMDS at St. Thomas's Hospital Medical School with H. S. Bachelard, and is supported by the Wellcome Trust; he is also Historian to the ISN.

Conventions

The abbreviations, conventions and symbols used in these Essays are those specified by the Editorial Board of *The Biochemical Journal* in *Policy of the Journal and Instructions to Authors* (see first issue in latest calendar year). The following abbreviations of compounds, etc., are allowed without definition in the text.

ADP, CDP, GDP, IDP, UDP, XDP, dTDP: 5'-pyrophosphates of adenosine, cytidine, guanosine, inosine, uridine, xanthosine and thymidine.

AMP, etc.: adenosine 5'-phosphate, etc.

ATP, etc.: adenosine 5'-triphosphate, etc.

CM-cellulose: carboxymethylcellulose

CoA and acyl-CoA: coenzyme A and its acyl derivatives

Cyclic AMP, etc.: adenosine 3',5'-cyclic phosphate, etc.

DEAE-cellulose: diethylaminoethylcellulose

DNA: deoxyribonucleic acid

Dnp-: 2,4-dinitrophenyl-

Dns-: 5-dimethylaminonaphthalene-1-sulphonyl-

EDTA: ethylenediaminetetra-acetate

FAD: flavin adenine dinucleotide

FMN: flavin mononucleotide

GSH, GSSG: glutathione, reduced and oxidized

NAD: nicotinamide adenine dinucleotide

NADP: nicotinamide adenine dinucleotide phosphate

NMN: nicotinamide mononucleotide

P_i, PP_i: orthophosphate, pyrophosphate

RNA: ribonucleic acid (see overleaf)

TEAE-cellulose: triethylammonioethylcellulose

tris: 2-amino-2-hydroxymethylpropane-1,3-diol

The combination NAD^+, NADH is preferred.

The following abbreviations for amino acids and sugars, for use only in presenting sequences and in Tables and Figures, are also allowed without definition.

Amino acids

Ala: alanine

Arg: arginine

Asn: asparagine

Asp: aspartic acid

Asx: aspartic acid or asparagine (undefined)

Cys: cystine or cysteine (half)

Gln: glutamine

Glu: glutamic acid

Glx: glutamic acid or glutamine (undefined)	Ile: isoleucine	Pro: proline
	Leu: leucine	Ser: serine
Gly: glycine	Lys: lysine	Thr: threonine
His: histidine	Met: methionine	Trp: tryptophan
Hyl: hydroxylysine	Orn: ornithine	Tyr: tyrosine
Hyp: hydroxyproline	Phe: phenylalanine	Val: valine

Sugars

Ara: arabinose	Glc*: glucose
dRib: 2-deoxyribose	Man: mannose
Fru: fructose	Rib: ribose
Fuc: fucose	Xyl: xylose
Gal: galactose	

*Where unambiguous, G may be used.

Abbreviations for nucleic acids used in these essays are:

mRNA: messenger RNA
nRNA: nuclear RNA
rRNA: ribosomal RNA
tRNA: transfer RNA

Other abbreviations are given on the first page of the text, or at first mention.

References are given in the form used in *The Biochemical Journal*, the last as well as the first page of each article being cited, and, in addition, the title. Titles of journals are abbreviated in accordance with the system employed in the *Chemical Abstracts Service Source Index* (1979) and its Quarterly Supplement (American Chemical Society).

Enzyme Nomenclature

At the first mention of each enzyme in each Essay there is given, whenever possible, the number assigned to it in *Enzyme Nomenclature: Recommendations (1984) of the Nomenclature Committee of the International Union of Biochemistry on the Nomenclature and Classification of Enzyme-catalysed Reactions*, published for the International Union of Biochemistry by Academic Press, New York and London, 1979. Enzyme numbers are given in the form EC 1.2.3.4. The names used by authors of the Essays are not necessarily those recommended by the International Union of Biochemistry.

Contents

BIOGRAPHY   v

CONVENTIONS   vii

PREFACE   xi

α-Lactalbumin and Related Proteins: A Versatile Gene Family with
an Interesting Parentage. By L. HALL and P. N. CAMPBELL . . .   1

Activation and Control of the Complement System. By K. B. M. REID   27

Processing and Metabolism of Neuropeptides. By A. J. TURNER . .   69

Fluorescence Analysis of Protein Dynamics. By A. P. DEMCHENKO .   120

The Origin and Use of the Terms Competitive and Non-competitive
in Interactions among Chemical Substances in Biological Systems.
By H. McILWAIN   158

SUBJECT INDEX   187

CUMULATIVE KEY WORD INDEX   197

Professor Rodney Porter O.M., F.R.S., Nobel Laureate
1917–1985

Preface

Peter Campbell was joint editor for the first twenty volumes of *Essays in Biochemistry* and as a parting gift he agreed to write a chapter for the present volume. This chapter revisits a topic that was dealt with in an earlier volume of this series (Brew, K. (1970): "Lactose synthetase: evolutionary origins, structure and control" *Essays in Biochemistry* **6**, 93–118). Comparison of the earlier chapter, which was contributed by a member of Peter Campbell's Department at the University of Leeds, with the material presented by Hall and Campbell in the present volume shows how the subject has developed in the past 16 years and particularly how the development of new techniques and approaches has fundamentally altered the nature of the questions that can be asked about a system such as this.

The sad death of Professor Rodney Porter, a recent photograph of whom appears opposite, is also commemorated in this issue in a chapter dedicated to his memory by Dr K. B. M. Reid, who worked with Rodney Porter at Oxford. Professor Porter's fundamental contributions to our understanding of the structure of the immunoglobins, for which he was awarded the Nobel Prize in 1972, are well known and it is particularly appropriate to have a contribution that involves some of his last work.

The other topics in this volume cover widely different fields. Our knowledge of peptides as neurotransmitters is still of relatively recent origin. The contribution by Dr A. J. Turner provides an account of the present state of understanding in this area which shows not only how rapidly our knowledge has grown, but also that there is still a great deal to be resolved. Dr A. P. Demchenko has made important contributions to the study of proteins by spectroscopic techniques. His chapter demonstrates how fluorescence studies have evolved from methods for the relatively gross assessment of protein conformation to a fine tool for assessing the dynamic structure of proteins and their interactions with ligands.

The final chapter in this volume is rather different in style and content. Professor Henry McIlwain is well known for his fundamental contributions in the area of neurochemistry. However, he has chosen to contribute a chapter on an entirely different topic: the development of ideas on reversible inhibition and their relationship with concepts outside the field of biochemistry. It is to be hoped that this chapter will be appreciated as an example of lateral thinking around a time-honoured topic.

This volume of *Essays in Biochemistry* contains five chapters and it is intended that future volumes will each contain five or six chapters in order to increase the coverage. Suggestions for topics to be included or potential authors would be most welcome.

R. D. Marshall & K. F. Tipton
June 1986

α-Lactalbumin and Related Proteins: A Versatile Gene Family with an Interesting Parentage

L. HALL[1] and P. N. CAMPBELL[2]

[1] Department of Biochemistry, University of Bristol, Bristol BS8 1TD, England
[2] Courtauld Institute of Biochemistry, The Middlesex Hospital Medical School, London W1P 7PN, England

I. Introduction 1
II. Structural Studies and Comparison between α-Lactalbumin and Lysozyme 4
III. Ion binding and the Interaction of α-Lactalbumin with Galactosyl Transferase 5
IV. The Hormonal Regulation of Milk Protein Synthesis . . 7
V. Structure of the α-Lactalbumin Gene from Different Species 11
VI. Comparison of the α-Lactalbumin Gene with that of Chick Lysozyme 15
VII. α-Lactalbumin and Breast Cancer 16
VIII. Identification of an α-Lactalbumin-like Protein in the Male Reproductive Tract 18
IX. Possible Function of the Epididymal α-Lactalbumin-like Activity 19
X. Conclusion and Future Research 21
Acknowledgements 22
References 23

I. Introduction

Our interest in α-lactalbumin arose when in 1964 one of us (P.N.C.), together with Keith Brew, embarked on a project to study the control of the biosynthesis of milk proteins as a model system. Milk proteins had for a long time held an attraction in this respect, for their synthesis is known to be controlled by several hormones. Guinea-pigs were chosen as the species of animal since they had discrete glands which were easily milked. Moreover, earlier experiments had shown that extracts, active in some steps of protein synthesis, could be prepared. Not wishing to complicate the experimental approach by following the synthesis of caseins, a complex family of phosphoproteins, we decided to study the synthesis of whey protein. At that

ESSAYS IN BIOCHEMISTRY Vol. 22
ISBN 0 12 158122 5

time, earlier experiments led us to believe that the predominant whey protein of the milk of all species, including guinea-pig, was β-lactoglobulin but it did not take long to show that virtually the sole whey protein in guinea-pig milk was α-lactalbumin. We had no alternative, therefore, but to study the biosynthesis of this protein if we were to stay with guinea-pigs. (It has subsequently been shown that in general β-lactoglobulin is present only in the milk of ruminants.) Guinea-pig α-lactalbumin proved to be a good choice for it is a comparatively small single chain protein which is not glycosylated and is comparatively easy to purify.

During our initial studies on α-lactalbumin one of us (P.N.C.) met Sir David Phillips who then was working at The Royal Institution on the structure of egg white lysozyme. We tried to interest Phillips in our protein, which for us had such excellent properties. He was unimpressed, regarding it as rather boring since it had no known function apart from serving as a source of nutrient for the young guinea-pig. This spurred us on to try to find a function and strangely many of the properties of α-lactalbumin transpired to be similar to lysozyme. Against some editorial opposition we were allowed to mention in one of two publications[1,2] that "the α-lactalbumin may have evolved by gradual modification from lysozyme, which is found in the milk of many species". Brew then moved to the laboratory of R. L. Hill and showed from elucidation of the primary structure that the two proteins had marked similarities (see Fig. 1).

The initial breakthrough in our understanding of the physiological function of α-lactalbumin is due to Brodbeck & Ebner[3] who were studying lactose synthase and showed that the enzyme could be resolved by gel filtration into two components, one of which, the so-called B protein, was identical to α-lactalbumin.[4] Brew, Vanaman & Hill[5] showed that the A protein catalysed another reaction:

UDP-d-galactose + N-acetyl d-glucosamine $\rightarrow$

$$UDP + N\text{-acetyllactosamine} \tag{1}$$

In the presence of increasing concentrations of α-lactalbumin the above reaction is inhibited and the following reaction is enhanced:

$$UDP\text{-galactose} + glucose \rightarrow lactose + UDP \tag{2}$$

The net effect of the B protein is to change the substrate (acceptor) specificity of the system from N-acetylglucosamine to glucose. Brew[6] in a previous volume of *Essays* postulated that the enzyme catalysing reaction

---

Fig. 1. A diagrammatic comparison of the covalent structure of human α-lactalbumin and hen's egg white lysozyme.

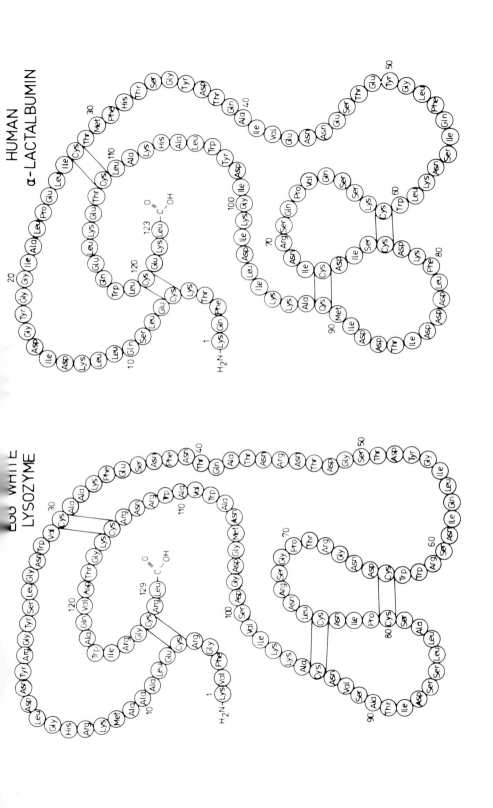

(1), *N*-acetyllactosamine synthase (galactosyl transferase), which is located in the Golgi Complex of most cells and is important in the biosynthesis of glycoproteins, interacts with α-lactalbumin produced in the rough-surfaced endoplasmic reticulum at the onset of lactation, leading to the synthesis of lactose.

It will be apparent that the above results, showing that α-lactalbumin has indeed an important physiological role, raised many interesting matters. Now seems an appropriate time to review recent progress. A computer search of papers mentioning α-lactalbumin that have been published over the last five years indicated a total of 109. These cover many aspects including purification and crystallization, which is an obvious preliminary to a study of the primary and tertiary structure; in particular it was interesting to compare the tertiary structures of α-lactalbumin and lysozyme. The role of α-lactalbumin as a so-called specifier protein which changes the affinity of an enzyme has also raised much interest, as has its ability to interact with the membranes of the endoplasmic reticulum and indeed model membranes also. Another rather more unexpected development has been the role of α-lactalbumin in ion binding, especially calcium; obviously an important matter in the composition of milk. It has been mentioned that α-lactalbumin is synthesized on the endoplasmic reticulum of the epithelial cells of the mammary gland and that this synthesis is under hormonal control. It has proved comparatively easy to isolate mRNA from the lactating mammary gland and to translate it *in vitro* to produce all the milk proteins, including α-lactalbumin. Not all such milk proteins are in the mature form as found in the milk. By this and other means some progress has been made in understanding the control of synthesis of α-lactalbumin and the caseins. A fuller understanding will depend on an analysis of the genes for the various proteins and of the regions in the genome on either side of the genes which no doubt control gene expression. Substantial progress has been made in this area which has depended on the use of recombinant DNA techniques.

Finally, an interesting twist to the subject has arisen from the discovery of an α-lactalbumin-like protein in the male reproductive tract.

II. Structural Studies and Comparison between α-Lactalbumin and Lysozyme

Since α-lactalbumin is virtually ubiquitous in the milk of all species, its primary structure has been a fruitful area of study. The only real problem has been to obtain reasonable quantities of protein from some of the smaller animals. As expected, there are substantial variations but with one exception the homologous proteins are of virtually the same size. The exception is rat α-lactalbumin which has a 17-residue-long carboxyl terminal extension, the sequence of which has been determined.[7] This protein, unlike most of the others, is also glycosylated, which accounts for its existence in three forms. Several possibilities exist to explain the carboxyl terminal extension, the most likely of which is a mutation of the expected termination codon.

Studies on the tertiary structure have been frustrated by the difficulty of crystallizing α-lactalbumin in a form suitable for X-ray crystallography. Several crystal forms have been obtained but with the exception of an orthorhombic crystal form of baboon α-lactalbumin none has proved suitable.[8] Studies of these crystals are now nearing completion (K. R. Acharya, D. C. Phillips, D. I. Stuart & N. P. C. Walker, personal communication). In addition to revealing the site of Ca^{2+} binding they show that the three-dimensional structure of α-lactalbumin is indeed very similar to lysozyme, as was suggested by earlier model building studies.[9] This finding supports the contention that the two proteins have been derived from a common ancestral gene. As will be shown later, this view has been fully supported by the recent studies on gene structure. It has not, of course, escaped attention that α-lactalbumin is involved in the formation of a $\beta1\rightarrow4$ disaccharide, whereas lysozyme hydrolyses a $\beta1\rightarrow4$ bond between N-acetylmuramic acid and N-acetylglucosamine. A knowledge of the tertiary structure of α-lactalbumin at high resolution and a detailed comparison with lysozyme should therefore provide some interesting insights into the mode of action of the former.

Other methods have been used to study the conformation of α-lactalbumin, particularly nuclear magnetic resonance and fluorescence techniques, but these studies involve ion binding and the interaction of α-lactalbumin with galactosyl transferase.

III. Ion Binding and the Interaction of α-Lactalbumin with Galactosyl Transferase

The nature of the interaction of the two proteins making up lactose synthase depends on an understanding of the conformation of the two proteins in the native state. While, as has already been mentioned, there have been difficulties in determining the tertiary structure of α-lactalbumin, the A protein, galactosyl transferase, has proved even more intractable, partly because of its size, M_r about 50 000, but mainly because of the difficulty of releasing it from the membranes of the endoplasmic reticulum in a soluble form.

A major development was the realization that α-lactalbumin is a calcium metalloprotein.[10] Highly purified α-lactalbumin contains a stoicheiometric quantity of calcium ions, namely one Ca^{2+} per molecule of protein. Moreover, removal of this Ca^{2+} by chelation with EDTA reduces the stability of the protein as judged by denaturation by heat or guanidine hydrochloride. This is consistent with the conformational change that takes place on cation binding.[11] It is interesting that lysozyme has only a low affinity for Ca^{2+}. (A comparison between the transient folding intermediates in lysozyme and α-lactalbumin has aroused much interest in view of the similar structure of the two proteins and their difference in function, see e.g. Kuwajima et al.[12]). Mn^{2+} also binds to the main Ca^{2+} binding site of α-lactalbumin, but with a lower affinity.

With reference to the binding of α-lactalbumin to A protein, Lindahl & Vogel,[13] using phenyl-Sepharose affinity chromatography, have suggested that a hydrophobic surface is exposed in apo-α-lactalbumin that disappears on calcium binding, in contrast to that of other calcium binding proteins such as calmodulin. They used this property to purify selectively apo-α-lactalbumin from Ca^{2+}-α-lactalbumin and postulated that this apolar surface on the protein may be physiologically important in its binding to A protein.

Berliner and his colleagues have intensively studied the binding of ions to α-lactalbumin and the effect of these on the conformation of the protein.[14] More recently they have carried out fluorescence studies with the dimeric form of the useful fluorophore bis-ANS.*[15] The conclusion is that Zn^{2+} binds to a site which differs from the Ca^{2+} binding site. When Zn^{2+} binds to the Ca^{2+}-protein the conformation is changed to an "apo-like" conformation that is not identical with that of the true apo form. Figure 2 depicts the four α-lactalbumin conformers and the equilibria which connect them.[16]

The question arises as to the role of the ions and the various conformers in lactose biosynthesis. For this the rate of production of UDP in reaction (2) was followed in the presence of Mn^{2+}. Musci and Berliner[16] thus showed that the K_m for lactose biosynthesis is the same in the presence of either the apo protein or the Ca^{2+}-bound protein. There was, however, a significant difference in the catalytic rate. They calculated that the apoprotein has a V_{max} which is 3·5 times higher than that of the Ca^{2+} protein. In the presence of Zn^{2+}, which shifts the protein towards an "apo-like" conformation, the K_m was unaffected and the V_{max} was as for the apoprotein.

Knowing the physiological concentration of Ca^{2+}, it is predicted that α-lactalbumin must be in the calcium form under physiological conditions and yet this form is less active in lactose synthesis. Calcium does, however, stabilize the protein against thermal unfolding. This would appear to be the role for calcium, whereas Zn^{2+} shifts the protein towards the "apo-like" conformation.

---

* bis-ANS:4,4'-bis[1-(phenylamino)-8-naphthalenesulphonate].

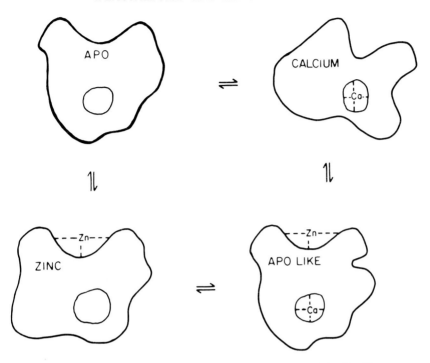

Fig. 2. Conformational states of α-lactalbumin (from Musci & Berliner[16]).

Assuming that free Zn^{2+} concentrations in the Golgi lumen are similar to those in milk, about 50 μM, α-lactalbumin should be predominantly in the "apo-like" conformer *in vivo*. It has also been suggested that Ca^{2+} might release and transport the newly synthesized protein from the endoplasmic reticulum to the Golgi lumen.[17] If α-lactalbumin binding to the membranes is analogous to its binding to phenyl-Sepharose or bis-ANS, only stoicheiometric amounts of Ca^{2+} would be necessary to effect release. In summary, Musci & Berliner[16] concluded that a balance between calcium and zinc concentrations "fine tunes" the protein conformation, modulating both α-lactalbumin release and its modifier activity in lactose biosynthesis.

IV. The Hormonal Regulation of Milk Protein Synthesis

The wide interest for biochemists and endocrinologists in the biosynthesis of milk proteins arises in part because synthesis of these substances is controlled by hormones. Hopefully we will eventually understand the

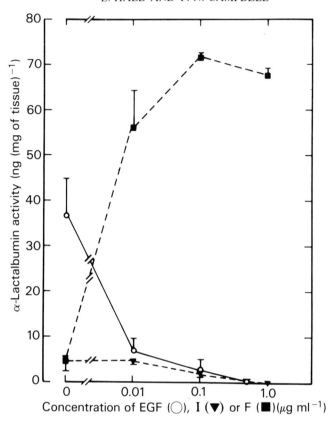

Fig. 3. Inhibition of prolactin-induced accumulation of α-lactalbumin activity as a function of EGF concentration: ability of cortisol and inability of insulin to reverse the inhibition. Mammary-gland explants were cultured for 4 days in medium containing prolactin (1 μg ml⁻¹) and various concentrations of EGF (○). Other explants were cultured for 4 days in medium containing prolactin (1 μg ml⁻¹) and a highly inhibitory concentration of EGF (0·1 μg ml⁻¹) to which various concentrations of insulin (I, ▼) or cortisol (F, ■) were added. Values for the EGF titration curve are the means ± SEM for eight rabbits, each analysed separately. Values for the insulin and cortisol titration curves are the mean ± range for two rabbits, each analysed individually (from ref. 19).

precise way in which the expression of the genes for the proteins is controlled.

In general terms, Topper and his co-workers,[18] from their work on the induction of casein and whey proteins in pregnant mouse mammary gland, showed that the effect of hormones could be divided into two stages: (a) a phase of cell division and differentiation invoked by the action of both insulin and hydrocortisone; and (b) a phase of induction which required the presence of insulin, hydrocortisone and prolactin. Progesterone, which is present during pregnancy but decreases in level at the onset of lactation, in general counteracts the action of prolactin. It is seen, therefore, that the hormonal regulation of milk proteins, at any rate in the mouse and rat, involves the synergistic and antagonistic interaction of at least four hormones (see Brew[6]).

The induction of milk proteins in the rabbit would appear to be simpler than in the mouse and rat. Sankaran & Topper[19] have studied the minimal hormone requirements in the rabbit for the induction of both casein and α-lactalbumin. They showed that, although prolactin induced α-lactalbumin when it is the only hormone added to the culture medium, at least two additional hormones are implicated physiologically in the response. Thus, as shown in Fig. 3, the prolactin-induced α-lactalbumin activity is profoundly inhibited by epidermal growth factor (EGF). Since EGF is a normal component of serum and its level is elevated during pregnancy, the effect of EGF has to be counteracted and cortisol would appear to play this role. Nicholas & Tyndale-Biscoe[20] have evidence that in the tammar, a marsupial, α-lactalbumin is induced maximally in explants by prolactin alone, and this was not inhibited by progesterone.

The biochemist who is looking for a beautifully simple explanation of events is in for another disappointment in that the expression of the milk protein genes differ temporally between species. Thus in the rat[21,22] and the mouse[23] casein and α-lactalbumin are produced throughout pregnancy, while the guinea-pig produces α-lactalbumin only two days before parturition and caseins only after parturition (see Fig. 4 and ref. 24). Recombinant plasmids containing DNA sequences complementary to the mRNAs for α-lactalbumin and each of three caseins[25] were used as probes. In respect of the caseins the level of mRNA was very low in the pregnant gland until 48 h before parturition, when there was a massive accumulation of casein mRNA transcripts. In the case of α-lactalbumin the mid-pregnant gland contained mRNA (62 days pregnant) and between this stage of development and parturition the amount increased 100-fold (Fig. 5). After parturition there was a small but significant decrease in α-lactalbumin mRNA. Explants of mammary tissue were used to examine the proteins synthesized at various stages of pregnancy. The pregnant gland synthesized α-lactalbumin, but translational control mechanisms appeared to regulate the onset of casein synthesis at parturition. Thus, in spite of the presence at parturition of mRNA for the caseins, only α-lactalbumin was being synthesized. This work also showed that different milk proteins are secreted at different rates within the same tissue. Thus in tissue from animals 72 h post-partum, two of the caseins were secreted faster than the third (Fig. 4). This work showing the differential synthesis of milk proteins in the guinea-pig makes it a particularly

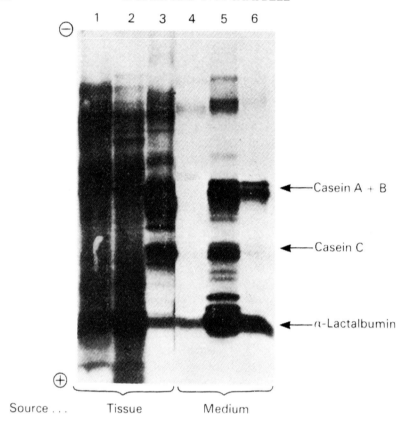

Fig. 4. Analysis of milk protein synthesis by explants of mammary tissue in culture, taken from guinea-pigs at parturition or 10 h and 72 h post-partum. ^{35}S-Methionine-labelled polypeptides were analysed by a combination of polyacrylamide-gel electrophoresis and antibody precipitation using antisera raised against guinea-pig milk proteins. Samples were as follows: total [^{35}S]-labelled polypeptides present in tissue (tracks 1, 2 and 3) or secreted into medium (tracks 4, 5 and 6). Tracks 1 and 4, parturition tissue; tracks 2 and 5, 10 h post-partum; tracks 3 and 6, 72 h post-partum (from ref. 24).

interesting species for studies on the effect of the various hormones on the expression of a small family of genes.

Returning to the effect of prolactin and hydrocortisone, the results of the early experiments certainly implied that prolactin was the major regulatory hormone.[26,27] Later experiments using organ culture showed that residual amounts of hydrocortisone remain in explants and that hydrocortisone is essential in producing a synergistic effect with prolactin.[28] Inhibition of hydrocortisone binding to the nuclear membrane has also been shown to

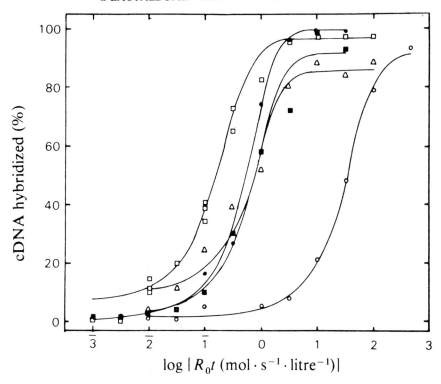

Fig. 5. Comparison of the relative amounts of α-lactalbumin RNA transcripts within the guinea-pig mammary gland during mammary development pre- and post-partum. Total RNA isolated from guinea-pig mammary tissue from animals 62 days pregnant (○), in premature labour (△), at parturition (□), and 8 h (●) and 7 days (■) post-partum, were hybridized in RNA excess to an α-lactalbumin sequence-specific cDNA probe (from ref. 24).

inhibit transcription of casein mRNA,[29] confirming the essential role of this hormone. The only precise role that has been demonstrated for prolactin is the stabilization of casein mRNA.[30] While hydrocortisone is essential for the induction of the caseins in mouse mammary explants, the optimum concentration is 3 μM. This is inhibitory for the synthesis of α-lactalbumin, which is induced at an optimum concentration of 30 nM.[31] Thus the differential regulation of caseins and α-lactalbumin in guinea-pigs may reside in the effect of hydrocortisone.

Thyroid hormones may also influence both the level of mRNA for α-lactalbumin and its rate of secretion in mouse mammary gland explants. Bhattacharjee & Vonderhaar[32] have recently shown that tri-iodo-thyronine selectively, with respect to α-lactalbumin, overcomes the inhibitory effect of hydrocortisone on the secretion of casein and α-lactalbumin.

V. Structure of the α-Lactalbumin Gene from Different Species

With the advent of recombinant DNA technology, and a realization of its potential, the trend in the 1970s was to clone and isolate specific cDNA probes, so that the control of hormonally or developmentally regulated genes could be studied at the transcriptional rather than at the protein level. In this respect α-lactalbumin was no exception and cDNA clones, and subsequently mRNA sequences, of rat,[33] guinea-pig[25,34] and human[34,35] α-lactalbumin were soon available. To a large extent studies with these cDNA probes confirmed earlier work on the differential control of α-lactalbumin and casein expression during mammary gland development, but in addition suggested that, for prolactin at least, control may be exerted at both transcriptional and post-transcriptional levels. In addition, the availability of specific, cloned cDNA sequences, made it a relatively simple task to isolate the α-lactalbumin genes from suitable genomic libraries, and the complete DNA sequences of the rat,[36] guinea-pig (J. E. Laird, L. Hall and R. K. Craig, unpublished observations) and human[37] α-lactalbumin genes, together with a substantial amount of their flanking sequences, is now known. Consequently, we can extend our already detailed knowledge of the evolution of α-lactalbumin protein sequences obtained from many, diverse species, to an analysis at the genomic DNA level.

All of the above three α-lactalbumin genes are composed of four exons separated by three introns. Furthermore, a study of the exon–intron boundaries reveals that the introns occur at identical positions in all three species; namely within amino acid residues Leu-26 (intron 1), Lys-79 (intron 2) and Trp-104 (intron 3) in the case of the human gene. However, although the size of the exons is highly conserved in these three genes, the size of the introns is not (see Table 1). Closer examination reveals that this is partly due to the

TABLE 1

Comparison of exon and intron lengths (in base pairs) of the human, guinea-pig and rat α-lactalbumin and chick lysozyme genes

Gene	Exon 1	Intron 1	Exon 2	Intron 2	Exon 3	Intron 3	Exon 4
Human α-lactalbumin	159	648	159	489	76	499	333
Guinea-pig α-lactalbumin	165	335	159	481	76	507	314
Rat α-lactalbumin	165	341	159	429	76	1016	328
Chick lysozyme	165	1270	162	1810	79	79	180

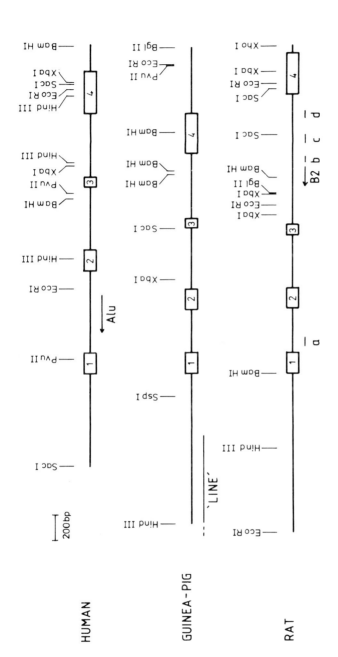

Fig. 6. Genomic organization of the human, guinea-pig and rat α-lactalbumin genes. Boxes indicate exons (1–4). The cleavage sites for a number of common hexanucleotide-specific restriction endonucleases are shown. "Alu", "LINE" and "B2" indicate the positions of repetitive sequences found within or adjacent to α-lactalbumin genes (see text). a–d represent repeating di- or trinucleotides found within the rat gene: a, (TCC)$_{23}$; b, (TG)$_{24}$; c, (TAT)$_{18}$; d, (TG)$_{21}$.

insertion of an Alu repetitive sequence (294 bp in length) within the first intron of the human α-lactalbumin gene (see Fig. 6). Deletion of this sequence would leave an intron of 354 bp, comparable in length to the first intron in the guinea-pig and rat genes, suggesting that the aquisition of this repetitive sequence occurred subsequent to the divergence of these species. Similar analysis of the rat α-lactalbumin gene reveals that intron 3 contains a B2 type repetitive sequence related to the human Alu repetitive family which is 160 bp in length.[39] As yet no function has been assigned to this type of repetitive sequence, although many have been shown to be transcribed by RNA polymerase III.[40] In addition, the rat sequence also differs from the human and guinea-pig genes in that introns 1 and 3 contain extended regions of regularly repeating dinucleotides (TG) or trinucleotides (TCC or TAT).

The guinea-pig α-lactalbumin gene is unique within this group of three, in

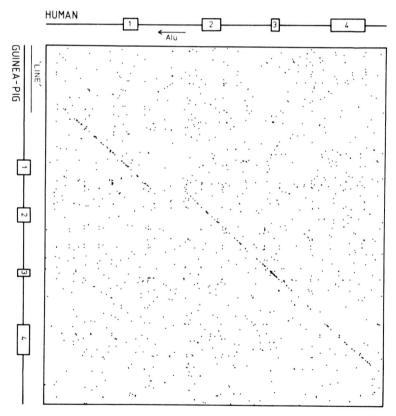

Fig. 7. DIAGON analysis of the human and guinea-pig α-lactalbumin genes. Areas of homology between the human and guinea-pig α-lactalbumin genes are indicated by means of a DIAGON plot (ref. 38) (parameters: span length = 7, score = 7).

that there do not appear to be any repetitive sequences associated with the introns. However, Southern blot analysis of the guinea-pig gene indicates the presence of a moderately repetitive sequence within the 5′ flanking region. A DNA database search subsequently revealed a significant degree of homology between the region ending about 500 bp upstream of the transcriptional initiation site, and other long interspersed elements ("Lines") found in primates and mice.[41]

As expected, DNA sequence comparison of any two of these three α-lactalbumin genes reveals a high degree of homology within the exons but a lesser homology in introns (see Fig. 7). The homology extends into the 5′ flanking region as far as the end of the long interspersed element in the case of the guinea-pig gene. In particular, there are several highly conserved regions found within the 5′ flanking region of all three genes[37] and these could be involved in one or more aspects of their hormonal control or tissue-specific expression. However, a definitive demonstration of the involvement of such regions in hormonal control must await the development of a mammalian system capable of supporting the hormonal regulation of reintroduced milk protein genes. It should then be possible, using site-directed mutagenesis, to establish precisely which regions are essential for control.

VI. Comparison of the α-Lactalbumin Gene with that of Chick Lysozyme

As stated earlier, it has been known for many years that α-lactalbumin and lysozyme share a significant degree of amino acid homology (about 38% for human α-lactalbumin versus chick lysozyme), suggesting evolution from a common ancestral gene. More recently, comparisons at the mRNA level[34] have revealed greater nucleotide than amino acid homology. Furthermore, of the codons which differed by only a single base, more than half represented silent substitutions. These observations strengthen the view that α-lactalbumin and lysozyme have arisen as a result of divergent evolution from a common gene, rather than convergent evolution from two quite distinct genes.

It therefore comes as no real surprise to find that the α-lactalbumin (human, guinea-pig and rat) and lysozyme (chick) genes have an identical exon–intron organization (Table 1), with introns occurring at homologous positions in the genes for the two proteins. In addition, it strongly suggests that the acquisition of introns preceded the gene duplication event which ultimately gave rise to α-lactalbumin and lysozyme. However, the wide variation in intron lengths between these two genes indicates major divergence within these regions, following gene duplication.

One popular theory for the existence of split genes is that exons originally encoded functional domains which recombined to form proteins of greater specificity or different function (see, for example, Cornish-Bowden[42]). It is therefore of interest to examine the exon structure of the α-lactalbumin and lysozyme genes in relation to their functional regions. The first three exons of the two genes show the highest degree of nucleotide and amino acid sequence homology, the fourth exon showing little, if any, similarity. In fact it is in this region of α-lactalbumin (from about residue 108) that the proposed conformation of the polypeptide backbone (based on model building) differs from that of lysozyme.[9] This may be explained to some extent by the proposal that a number of residues within exon 4 play an important role in the interaction of α-lactalbumin with galactosyl trans-ferase[43,44]—a function which is not required by lysozyme.

Exon 2 appears to be the most highly conserved exon in these two genes. In lysozyme this region contains many of the residues implicated in oligo-saccharide binding. The conservation of such residues in α-lactalbumin suggests that they may also play a role in carbohydrate binding in the lactose–synthase complex. This exon also contains the two residues (Asp-53 and Glu-35) essential for the catalytic activity of lysozyme, but which differ in α-lactalbumin.

The two sets of paired dicarboxylic acid residues (Asp-82, -83, -87 and -88), which are conserved within exon 3 of α-lactalbumin, but not in lysozyme, have been shown by X-ray crystallography (K. R. Acharya, D. C. Phillips, D. I. Stuart & N. P. C. Walker, personal communication) to be involved in the Ca^{2+} binding site. Exon 3 of lysozyme, on the other hand, contains residues which confer additional substrate specificity, increase the catalytic activity and determine the specific cleavage frame of the alternating N-acetylglucosamine/N-acetylmuramic acid copolymer.

In conclusion, although α-lactalbumin and lysozyme do not ahdere rigidly to the exon-domain theory, some generalizations can probably be made. Exon 2 is the most highly conserved exon and appears to have a similar function in both proteins, that of substrate binding, although in lysozyme this exon also contains some catalytic site residues. Exon 4 plays an important role in galactosyl transferase binding in α-lactalbumin and hence shows very little homology with its lysozyme counterpart. Finally, the conserved aspartyl residues in exon 3 account for the high affinity of α-lactalbumin for calcium ions.

VII. α-Lactalbumin and Breast Cancer

About a third of all human metastatic breast carcinomas regress in response to some form of endocrine therapy;[45] yet, despite much research, there is still no reliable way of identifying this group prior to treatment. For many years determination of the hormone receptor status of tumour tissue has provided the only indication of potentially hormone-receptive tumours, but the correlation is far from good.

An alternative approach, which received much attention in the late 1970s, was to look for milk proteins within breast tumours or serum. The rationale behind this approach was based on the knowledge that lactation is clearly a hormonally regulated process and hence it could be argued that the expression of milk proteins might identify those tumours which are hormone responsive (see ref. 46). However, although substantiated to some extent in experimental animal systems,[47,48] results with human mammary carcinomas were inconclusive, mainly because of insufficient specificity (see Stevens *et al.*[49]) and sensitivity of the radioimmunological protein assay used to determine α-lactalbumin levels. It was mainly because of these technical problems that one of us (L.H.), in collaboration with Roger Craig, decided to assay α-lactalbumin expression at the transcriptional level, by measuring steady-state mRNA levels in breast tumours using a cloned α-lactalbumin-specific cDNA probe.[35] It had been already established that total mRNA isolated from some of the human breast tumours was capable of directing the synthesis, *in vitro*, of a peptide with an identical molecular weight to that of pre-α-lactalbumin,[50] and that this peptide could be specifically precipitated by antisera raised against highly purified human α-lactalbumin (see Fig. 8). It therefore came as something of a surprise when total mRNA preparations isolated from a variety of breast tumours were screened with the α-lactalbumin cDNA probe, and it was found that α-lactalbumin transcripts were not present in any of them[51] under conditions capable of detecting a single mRNA copy per cell. This was in spite of the fact that some of these tumour samples had been independently shown by radioimmunoassay to contain significant amounts of a substance assaying as "α-lactalbumin".

In conclusion, therefore, α-lactalbumin was not being expressed in any of the breast tumours examined, but instead it appeared that α-lactalbumin antisera was capable of reacting with a peptide which not only shared common antigenic determinants with α-lactalbumin, but was also of the same apparent molecular weight as pre-α-lactalbumin, when analysed on one-dimensional SDS-polyacrylamide gels. Subsequent two-dimensional gel analysis (L. Hall & R. K. Craig, unpublished data) has revealed that this peptide can in fact be resolved from pre-α-lactalbumin on the basis of its isoelectic point, thereby confirming its separate identity.

Clearly, α-lactalbumin has not fulfilled the expectation that it might be a useful marker of human hormone-dependent breast cancer. The precise nature of the peptide present in breast tumours, and precipitated by human α-lactalbumin antisera, has yet to be established. Such a task is certainly worthwhile since there is a good correlation between its presence, and the

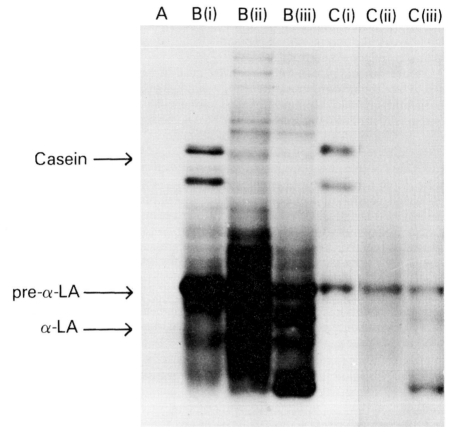

Fig. 8. Cell-free translation of poly A-containing RNA isolated from human breast samples (from ref. 52). mRNA activity was determined in a wheat-germ cell-free protein synthesizing system in the presence of [^{35}S]methionine. The products were electrophoresed on an SDS-polyacrylamide gel and visualized by fluorography. (A) Wheat-germ proteins synthesized in the absence of added RNA. (B) proteins synthesized on addition of poly A-containing RNA from (i) human lactating mammary tissue, (ii) and (iii) two different breast tumour samples. (C) Proteins as in (B) precipitated by combined antisera to human α-lactalbumin and the major β-casein. Arrows indicate the positions of authentic mature α-lactalbumin and β-casein as isolated from human milk, and the position of pre-α-lactalbumin synthesized in the cell-free system. The peptide which co-eletrophoresed with pre-α-lactalbumin in the breast tumour tracks was subsequently shown not to be pre-α-lactalbumin (see text).

presence of oestrogen receptors, making it a possible candidate as a marker of an intact oestrogen receptor mechanism, and hence possibly an indicator of hormone-responsive breast cancer. However, the relationship, if any, of this peptide to α-lactalbumin has yet to be established, since despite cross-reaction at the immunological level there appears to be little homology at the DNA level.

VIII. Identification of an α-Lactalbumin-like Protein in the Male Reproductive Tract

Sperm maturation, the process by which immature spermatoza achieve their full fertilizing capacity, is clearly an important aspect of mammalian reproductive biology. Although the subject of much debate, it is generally accepted that normal sperm maturation is a complex co-operative process in which sperm, epididymal fluid, epididymal epithelium and their interstitium all play important parts. For this reason the epididymis, in particular that of the rat, has been the focus of much attention and a great deal of effort has been directed towards characterizing the proteins synthesized by the epididymis and found in epididymal fluid, a number of which have been shown to be regulated by androgens.[52,53] In addition, the sperm surface glycoproteins have been the subject of much study[54,55] since their glycosylation patterns undergo considerable changes during passage through the epididymis, some changes showing a close correlation with the acquisition of fertilizing capacity. It was during the course of these studies that Professor D. Hamilton of the University of Minnesota found rat rete testis and epididymal fluids to be rich sources of galactosyl transferase activity.[56] Although many of the properties of this enzyme differed from those of the mammary gland and serum enzymes, it still possessed the ability to respond to purified α-lactalbumin from the mammary gland in the way characteristic of that tissue. Hamilton therefore decided to look for a protein with α-lactalbumin activity in rat epididymal fluid and one was indeed found to be present in reasonable abundance.[57] However, this protein differed from its mammary counterpart in that it promoted the transfer of UDP-galactose to inositol or glucose with equal efficiency, whereas only glucose can act as the acceptor in the presence of mammary gland α-lactalbumin.

The most recent work from Hamilton's group suggests that the purified epididymal fluid protein exhibiting most of the α-lactalbumin activity is different from its mammary counterpart, on the bases of both immunological studies and limited N-terminal protein sequence analysis, despite having a similar molecular weight.

We are therefore faced with the intriguing possibility that the α-lactalbumin/lysozyme gene family may possess a third member, which has provisionally been called α-lactalbumin-like protein (α-LAL) by Hamilton. Clearly, available data on the activity of this "new" protein would suggest it is likely to be more closely related to mammary gland α-lactalbumin than to lysozyme, and a complete analysis of its primary (and hopefully secondary and tertiary) structure should be particularly revealing.

IX. Possible Function of the Epididymal α-Lactalbumin-like Activity

At the moment, one can only speculate as to the function of an α-lactalbumin-like activity in the male reproductive tract. Clearly, lactose synthesis is not a major function for the putative "α-lactalbumin"/galactosyl transferase complex in the male, since neither testicular fluid nor epididymal fluid contain detectable levels of free glucose. However, the observation[57] that inositol, which is present in high amounts in these fluids, can act as an efficient acceptor for galactose makes it a prime candidate as a substrate. One possibility is that inositol in sperm membrane phosphatidylinositol might act as an acceptor for galactose in a manner similar to phosphatidylinositol mannoside synthesis in bacteria. However, perhaps a more likely (and exciting) function might be that it is involved in regulating the changing glycosylation patterns of sperm surface glycoproteins by sequential removal and addition of sugar residues, during passage through the epididymal tract. Such a mechanism would require not only an "α-lactalbumin"/galactosyl (or other glycosyl) transferase complex to catalyse the addition of monosaccharide residues, but also glycosidases for their removal. In fact, the epididymis is probably the richest source of glycosidases in the body and their secretion into epididymal fluid is regulated by androgens.[58] In support of such a mechanism, studies on the labelling of the carbohydrate moieties of externally orientated cell surface glycoproteins have revealed that as spermatozoa proceed through the epididymis there is a sequential change from the labelling of a 110 kd glycoprotein on testicular spermatozoa to a 32–37 kd glycoprotein on cauda epididymal spermatozoa, and this change in the glycosylation pattern occurs in the caput region,[54,59] the region associated with maximum α-lactalbumin-like activity. Interestingly, the appearance of the glycosylated 32–37 kd glycoprotein in the proximal cauda epididymis, coincides with the completion of the sperm maturation process.

The idea of sequential removal and addition of carbohydrate moieties is not without precedent, but bears many similarities to differentiation antigens—antigens whose expression on cell surfaces varies during development (reviewed by Feizi[60]). Recent analysis of some differentiation antigens, using monoclonal antibodies specific for defined carbohydrate sequences, have shown that some of the changing antigencies may be caused by sequential removal or addition of sugar residues from glycoprotein or glycolipids, or both. Consequently, in view of the marked decrease in galactosyl transferase activity in the distal cauda region of the epididymis, coincident with the appearance of the 32–37 kd surface glycoprotein, and the acquisition of sperm fertilizing ability, it is tempting to speculate that the precise glycosylation pattern of sperm surface glycoproteins, regulated by an "α-lactalbumin"/galactosyl transferase complex, may play an important part in sperm maturation, possibly at the level of sperm–egg recognition and subsequent fusion of cell membranes.

In fact Shur and Hall have already established that the initial binding of mouse sperm to the egg zona pellucida involves the sperm surface galactosyl transferase[61,62] and they have also shown that the addition of mammary α-lactalbumin prevents sperm–egg binding *in vitro*, in a dose dependent manner.[63]

X. Conclusion and Future Research

Peptide and steroid hormones play a major role in the regulation of tissue differentiation during development. One of the ways in which they achieve this is by modulating the expression of a number of gene products that are characteristic of a target cell. In some instances hormonal control appears to act primarily at the level of transcription, while in others, post-transcriptional mechanisms have been implicated. With respect to steroid hormones, it has recently been established that hormone–receptor complexes bind to specific sites within the promoter regions of some hormonally controlled genes and the primary structure of the human glucocorticoid receptor is known.[63] Analysis of the expression of these genes after their reintroduction into mammalian cells has demonstrated the importance of these binding sites for specific regulation of transcription. In contrast, little is known about the way in which peptide hormones modulate specific gene expression.

As a model system for such studies, α-lactalbumin, which is controlled by the peptide hormone prolactin, has much to offer. Our current understanding of the hormonal control of this protein is based on a combination of studies on the whole animal, mammary explants, organ culture, and isolated epithelial cells in culture. Unfortunately, it is now clear that such studies suffer from a number of limitations, which include problems of residual hormone levels, cell type heterogeneity, and the inability to manipulate directly specific gene sequences to establish which regions are essential for hormonal regulation. However, now that the α-lactalbumin genes have been isolated and characterized from a number of species, these problems can be overcome by studying the expression of the cloned genes after their reintroduction into hormone-responsive cell lines, both before and after manipulation of putative regulatory regions. This approach is currently being pursued by a number of groups and it is hoped that we should soon know precisely which hormones are involved in the control of α-lactalbumin expression and which regions of the genome are essential for this control. In the longer term, we should be able to identify and isolate the putative DNA binding component (since neither prolactin nor its receptor appear to be translocated to the nucleus), and ultimately establish the precise mode of action of prolactin in the regulation of α-lactalbumin gene expression.

But the interest in α-lactalbumin clearly does not stop there. It is not just a "model protein" for hormonal control; instead it is a member of a small gene family with very diverse properties. The discovery that α-lactalbumin and lysozyme are structurally related, having arisen from a common ancestral gene, has made them a particularly attractive system for studying protein structure–function relationships as well as various aspects of protein evolution. Here we have two proteins with very similar tertiary structures involved in two quite distinct enzymic reactions—one capable of modifying the substrate specificity of an enzyme involved in disaccharide synthesis, the other an enzyme itself, responsible for the hydrolysis of a $\beta 1 \rightarrow 4$ glycosidic linkage in the mucopolysaccharide component of some bacterial cell walls.

Recently the relationship between α-lactalbumin and lysozyme has given rise to speculation concerning the origin of lactation.[64] Attention has been drawn to the advantage to the survival of the eggs or young by virtue of a protein with antimicrobial properties.

The recent success from the X-ray crystallographers in determining the three-dimensional structure of baboon α-lactalbumin is particularly gratifying. Not only is the binding site for Ca^{2+} now understood in great detail, but this knowledge encourages further studies on the interaction between α-lactalbumin and galactosyl transferase to explain the way in which it acts to change the substrate specificity of that enzyme. Although the early prediction that the "specifier" role of α-lactalbumin would be only the first of many examples of this phenomenon has not so far been fulfilled, there are an increasing number of examples of the activation of an enzyme by another protein, particularly in the membrane receptor field. Future work on the interaction of the A and B proteins of lactose synthase may well have wider implications.

In addition there is now a strong possibility that the α-lactalbumin/lysozyme gene family may have a third member; the so-called α-lactalbumin-like protein (α-LAL) recently found in the mammalian male reproductive tract. α-LAL is clearly more akin to mammary gland α-lactalbumin than to lysozyme, at least in terms of function, since both can interact with galactosyl transferase to form a lactose synthase complex *in vitro*. However, the postulated function of the α-LAL protein *in vivo* in the galactosylation of sperm surface proteins must require some unique structural properties not found in the other two members of the family.

We have learnt a great deal over the past 20 years about this versatile family of proteins, but clearly there is still much work to be done. We hope that we have convinced our readers that we, and others, were right to pursue our studies on a protein which was regarded initially as rather boring.

ACKNOWLEDGEMENTS

The research work described in this Essay has clearly depended on workers in many laboratories and we would like to express our gratitude to them all. We would particularly like to express our thanks to our colleagues, Keith Brew for his work as a protein chemist and Roger Craig as an expert in the application of recombinant DNA techniques. They have together inspired much of the work. We are also grateful to Lawrence Berliner and to Sir David Phillip's group at Oxford for their co-operation.

REFERENCES

1. Brew, K. & Campbell, P. N. (1967). The characterization of the whey proteins of guinea-pig milk. *Biochem. J.* **102**, 258–264.
2. Brew, K. & Campbell, P. N. (1967). Studies on the biosynthesis of protein by lactating guinea-pig mammary gland. *Biochem. J.* **102**, 265–274.
3. Brodbeck, U. & Ebner, K. E. (1966). Resolution of a soluble lactose synthetase into two protein components and solubilization of microsomal lactose synthetase. *J. Biol. Chem.* **241**, 762–764.
4. Brodbeck, U., Denton, W. L., Tanahashi, N. & Ebner, K. E. (1967). The isolation and identification of the B protein of lactose synthetase as α-lactalbumin. *J. Biol. Chem.* **242**, 1391–1397.
5. Brew, K., Vanaman, T. C. & Hill, R. L. (1968). The role of α-lactalbumin and the A protein in lactose synthetase: a unique mechanism for the control of a biological reaction. *Proc. Natl. Acad. Sci. USA* **59**, 491–497.
6. Brew, K. (1970). Lactose synthetase: Evolutionary origins, structure and control. In *Essays in Biochemistry* (Campbell, P. N. & Dickens, F., eds) Vol. 6, pp. 93–118. Academic Press, London.
7. Prasad, R. V., Butkowski, R. J., Hamilton, J. W. & Ebner, K. E. (1982). Amino acid sequence of rat α-lactalbumin: a unique α-lactalbumin. *Biochemistry* **21**, 1479–1482.
8. Aschaffenburg, R., Fenna, R. E., Phillips, D. C., Smith, S. G., Buss, D. H., Jenness, R. & Thompson, M. P. (1979). *J. Mol. Biol.* **127**, 135–137.
9. Browne, W. J., North, A. C. T., Phillips, D. C., Brew, K., Vanaman, T. C. & Hill, R. L. (1969). A possible three-dimensional structure of bovine α-lactalbumin based on that of hen's egg-white lysozyme. *J. Mol. Biol.* **42**, 65–86.
10. Hiraoka, Y., Segawa, T., Kuwajima, K., Sugai, S. & Murai, N. (1980). α-lactalbumin: a calcium metalloprotein. *Biochem. Biophys. Res. Commun.* **93**, 1098–1104.
11. Permyakov, E. A., Yarmolenko, V. V., Kalinichenko, L. P., Morozova, L. A. & Burstein, E. A. (1981). Calcium binding to α-lactalbumin: structural rearrangement and protein fluorescence changes. *Biochem. Biophys. Res. Commun.* **100**, 191–197.

12. Kuwajima, K., Hiraoka, Y., Ikeguchi, M. & Sugai, S. (1985). Comparison of the transient folding intermediates in lysozyme and α-lactalbumin. *Biochemistry* **24**, 874–881.
13. Lindahl, L. & Vogel, H. J. (1984). Metal-ion-dependent hydrophobic-interaction chromatography of α-lactalbumins. *Anal. Biochem.* **140**, 394–402.
14. Murakami, K., Andree, P. J. & Berliner, L. J. (1982). Metal ion binding to α-lactalbumin species. *Biochemistry* **21**, 5488–5494.
15. Musci, G. & Berliner, L. J. (1985). Probing different conformational states of bovine α-lactalbumin: Fluorescence studies with 4,4′-Bis [1-(phenylamino)-8-naphthalenesulfonate]. *Biochemistry* **24**, 3852–3856.
16. Musci, G. & Berliner, L. J. (1985). Physiological roles of zinc and calcium binding to α-lactalbumin in lactose biosynthesis. *Biochemistry* **24**, 6945–6948.
17. Hill, R. L. & Brew, K. (1975). Lactose synthetase. *Adv. Enzymol.* **43**, 411–490.
18. Turkington, R. W., Lockwood, D. H. & Topper, Y. J. (1967). The induction of milk protein synthesis in post-mitotic mammary epithelial cells exposed to prolactin. *Biochim. Biophys. Acta* **148**, 475–480.
19. Sankaran, L. & Topper, Y. J. (1984). Prolactin-induced α-lactalbumin activity in mammary explants from pregnant rabbits. *Biochem. J.* **217**, 833–837.
20. Nicholas, K. R. & Tyndale-Biscoe, C. H. (1985). Prolactin-dependent accumulation of α-lactalbumin in mammary gland explants from the pregnant tammar wallaby (*Macropus eugenii*). *J. Endocrinol.* **106**, 336–342.
21. Rosen, J. M., Woo, S. L. C. & Comstock, J. P. (1975). Regulation of casein messenger RNA during development of the rat mammary gland. *Biochemistry* **14**, 2895–2903.
22. Rosen, J. M., O'Neil, D. L., McHugh, J. E. & Comstock, J. P. (1978). Progesterone-mediated inhibition of casein mRNA and polysomal casein synthesis in the rat mammary gland during pregnancy. *Biochemistry* **17**, 290–296.
23. Anderson, R. R. (1974). Endocrinological control in the development of the mammary gland. In *Lactation* (Larson, B. L. & Smith, V. R., eds) Vol. 1, pp. 97–140. Academic Press, New York.
24. Burditt, L. J., Parker, D., Craig, R. K., Getova, T. & Campbell, P. N. (1981). Differential expression of α-lactalbumin and casein genes during the onset of lactation in the guinea-pig mammary gland. *Biochem. J.* **194**, 999–1006.
25. Craig, R. K., Hall, L., Parker, D. & Campbell, P. N. (1981). The construction, identification and partial characterization of plasmids containing guinea-pig milk protein complementary DNA sequences. *Biochem. J.* **194**, 989–998.
26. Ono, M. & Oka, T. (1980). The differential actions of cortisol on the accumulation of α-lactalbumin and casein in midpregnant mouse mammary gland in culture. *Cell* **19**, 473–480.
27. Matusik, R. J. & Rosen, J. M. (1978). Prolactin induction of casein mRNA in organ culture. *J. Biol. Chem.* **253**, 2343–2347.
28. Mehta, N. M., Ganguly, M., Ganguly, R. & Banerjee, M. R. (1980). Hormonal modulation of the casein gene expression in a mammogenesis-lactogenesis culture model of the whole mammary gland of the mouse. *J. Biol. Chem.* **255**, 4430–4434.
29. Majunder, P. K., Joshi, J. B. & Banerjee, M. R. (1983). Correlation between nuclear glucocorticoid receptor levels and casein gene expression in murine mammary gland *in vitro*. *J. Biol. Chem.* **258**, 6793–6798.
30. Guyette, W. A., Matusik, R. J. & Rosen, J. M. (1979). Prolactin-mediated

transcriptional and post-transcriptional control of casein gene expression. *Cell* **17**, 1013–1023.

31. Ono, M. & Oka, T. (1980). α-lactalbumin-casein induction in virgin mouse mammary explants: Dose-dependent differential action of cortisol. *Science* **207**, 1367–1369.

32. Bhattacharjee, M. & Vonderhaar, B. K. (1985). Thyroid hormones enhance the synthesis and secretion of α-lactalbumin by mouse mammary tissue *in vitro*. *Endocrinology* **115**, 1070–1077.

33. Dandekar, A. M. & Qasba, P. K. (1981). Rat α-lactalbumin has a 17-residue-long COOH-terminal hydrophobic extension as judged by sequence analysis of the cDNA clones. *Proc. Natl. Acad. Sci. USA* **78**, 4853–4857.

34. Hall, L., Craig, R. K., Edbrooke, M. R. & Campbell, P. N. (1982). Comparison of the nucleotide sequence of cloned human and guinea-pig pre α-lactalbumin cDNA with that of chick pre-lysozyme cDNA suggests evolution from a common ancestral gene. *Nucleic Acids Res.* **10**, 3503–3515.

35. Hall, L., Davies, M. S. & Craig, R. K. (1981). The construction, identification and characterisation of plasmids containing human α-lactalbumin cDNA sequences. *Nucleic Acids Res.* **9**, 65–84.

36. Qasba, P. K. & Safaya, S. K. (1984). Similarity of the nucleotide sequences of rat α-lactalbumin and chicken lysozyme genes. *Nature* **308**, 377–380.

37. Hall, L., Emery, D. C., Davies, M. S., Parker, D. & Craig, R. K. (1986). Organisation and sequence of the human α-lactalbumin gene. *Biochem. J.* (in press).

38. Staden, R. (1982). An interactive graphics program comparing and aligning nucleic acid and amino acid sequences. *Nucleic acids Res.* **10**, 2951–2961.

39. Rogers, J. H. (1985). The origin and evolution of retroposons. *Int. Rev. Cytol.* **93**, 188–279.

40. Fuhrmann, S. A., Deininger, P. L., La Porte, P., Friedmann, T. & Geiduschek, E. P. (1981). Analysis of transcription of the human Alu family ubiquitous repeating element by eukaryotic RNA polymerase III. *Nucleic Acids Res.* **9**, 6439–6456.

41. Rogers, J. (1983). A straight LINE story. *Nature* **306**, 113–114.

42. Cornish-Bowden, A. (1985). Are introns structural elements or evolutionary debris? *Nature* **313**, 434–435.

43. Shewale, J. G., Sinha, S. K. & Brew, K. (1984). Evolution of α-lactalbumins. The complete amino acid sequence of α-lactalbumin from a marsupial (*Macropus rufogriseus*) and corrections to regions of sequence in bovine and goat α-lactalbumins. *J. Biol. Chem.* **259**, 4947–4956.

44. Jung, A., Sippel, A. E., Grez, M. & Schutz, G. (1980). Exons encode functional and structural units of chicken lysozyme. *Proc. Natl. Acad. Sci. USA* **77**, 5759–5763.

45. McGuire, W. L., Horwitz, K. B., Pearson, O. H. & Segaloff, A. (1977). Current status of estrogen and progesterone receptors in breast cancer. *Cancer* **39**, 2934–2947.

46. Woods, K. L., Cove, D. H. & Howell, A. (1977). Predictive classification of human breast carcinomas based on lactalbumin synthesis. *Lancet* **ii**, 14–16.

47. Ceriani, R. L., Contesso, G. P. & Nataf, B. M. (1972). Hormone requirement for growth differentiation of the human mammary gland in organ culture. *Cancer Res.* **32**, 2190–2196.

48. Rosen, J. M. & Socher, S. H. (1977). Detection of casein messenger RNA in hormone-dependent mammary cancer by molecular hybridisation. *Nature* **269**, 83–86.
49. Stevens, U., Laurence, D. J. R. & Ormerod, M. G. (1978). Antibodies to lactalbumin interfere with its radioimmunoassay in human plasma. *Clin. Chim. Acta* **87**, 149–157.
50. Hall, L., Davies, M. S. & Craig, R. K. (1981). mRNA species directing synthesis of milk proteins in normal and tumour tissue from mammary gland. *Nature* **277**, 54–56.
51. Hall, L., Craig, R. K., Davies, M. S., Ralphs, D. N. L. & Campbell, P. N., (1981). α-lactalbumin is not a marker of human hormone-dependent breast cancer. *Nature* **290**, 602–604.
52. D'Agostino, A., Jones, R., White, R. & Parker, M. G. (1980). Androgenic regulation of messenger RNA in rat epididymis. *Biochem. J.* **190**, 505–512.
53. Jones, R., Brown, C. R., von Glos, K. I. & Parker, M. G. (1980). Hormonal regulation of protein synthesis in the rat epididymis. Characterization of androgen-dependent and testicular fluid-dependent proteins. *Biochem. J.* **188**, 667–676.
54. Olson, G. E. & Hamilton, D. W. (1978). Characterization of the surface glycoproteins of rat spermatozoa. *Biol. Reprod.* **19**, 26–35.
55. Jones, R., von Glos, K. I. & Brown, C. C. (1981). Characterization of hormonally regulated secretory proteins from the caput epididymis of the rabbit. *Biochem. J.* **196**, 105–114.
56. Hamilton, D. W. (1980). UDP-galactose: N-acetylglucosamine galactosyl transferase in fluids from rat testis and epididymis. *Biol. Reprod.* **23**, 377–385.
57. Hamilton, D. W. (1981). Evidence for α-lactalbumin-like activity in reproductive tract fluids of the male rat. *Biol. Reprod.* **25**, 385–392.
58. Jones, R. (1974). Absorption and secretion in the cauda epididymis of the rabbit and effects of degenerating spermatozoa on epididymal plasma after castration. *J. Endocrinol.* **63**, 157–165.
59. Jones, R., Pholpramool, C., Setchell, B. P. & Brown, C. R. (1981). Labelling of membrane glycoproteins on rat spermatozoa collected from different regions of the epididymis. *Biochem. J.* **200**, 457–460.
60. Feizi, T. (1981). Carbohydrate differentiation antigens. *Trends Biochem. Sci.* **6**, 333–335.
61. Shur, B. D. & Hall, N. G. (1982). Sperm surface galactosyl transferase activities during *in vitro* capacitation. *J. Cell Biol.* **95**, 567–573.
62. Shur, B. D. & Hall, N. G. (1982). A role for mouse sperm surface galactosyl transferase in sperm binding to the egg zona pellucida. *J. Cell Biol.* **95**, 574–579.
63. Hollenberg, S. M., Weinberger, C., Ong, E. S., Cerelli, G., Oro, A., Lebo, R., Thompson, E. B., Rosenfeld, M. G. & Evans, R. M. (1985). Primary structure and expression of a functional human glucocorticoid receptor cDNA. *Nature* **318**, 635–641.
64. Hayssen, V. & Blackburn, D. G. (1985). α-lactalbumin and the origin of lactation. *Evolution* **39**, 1147–1149.

Activation and Control of the Complement System

KENNETH B. M. REID

*M.R.C. Immunochemistry Unit, Department of Biochemistry, University of Oxford,
South Parks Road, Oxford OX1 3QU, England*

I. Introduction	28
II. Activation of the Early-acting Components of the Classical Pathway of Complement Fixation	31
A. The C1 Complex and Activators of the Classical Pathway	31
B. Components C4 and C2 and Formation of the C3 Convertase C4b, 2a	36
C. Component C3	39
III. Activation of the Alternative Pathway	41
A. Formation of the Initial C3 Convertase	41
B. Factors D and B and Formation of the C3 Convertase C3b,Bb	42
C. Activators of the Alternative Pathway	42
IV. Activation of the Terminal Components C5 to C9	43
A. Formation of C5 Convertase Activity	43
B. Activation of C5 and Assembly of C5b–8	45
C. Component C9 and Membrane Lesion Formation	46
V. Control Proteins Present in Plasma	47
A. Control of Activated C1 by C1-inhibitor	47
B. C4b-binding Protein and Factor I in Regulation of C4b	48
C. Factor H and Factor I in Regulation of C3b	48
D. Anaphylatoxin Inactivator	49
E. S-Protein in the Control of Formation of the C5b-9 Complex	49
F. Properdin and Stabilization of C3b,Bb	50
VI. Membrane-associated Regulatory Proteins	50
A. The C3b/C4b Receptor (CR1)	50
B. Decay Accelerating Factor (DAF)	50
C. Glycoproteins 45–70 (gp 45–70)	51
VII. Molecular Cloning and Genetics	51
A. Introduction	51
B. Subcomponent C1q	52
C. Components C4, C2 and Factor B	54
D. C4b-binding Protein, Factor H and CR1	58
E. Structural Evidence that C2, Factor B, C4BP, Factor H and CR1 are Part of a Superfamily of Related Proteins	58
F. Components C3, C5, C9 and the Control Protein, S-Protein	61

ESSAYS IN BIOCHEMISTRY Vol. 22
ISBN 0 12 158122 5

VIII. Conclusion   62
      References   63

I. Introduction

This article is dedicated to the memory of Professor Rodney Porter, Companion of Honour, Fellow of the Royal Society, whose death in a car accident occurred on 6 September 1985. During the period 1955–67, Professor Porter's research was concerned mainly with structural studies on immunoglobulin molecules. This work included the solution of the well-known four-chain structure of the antibody IgG molecule for which he was awarded, jointly with G. M. Edelman, the Nobel Prize for Physiology or Medicine in 1972. As Director of the MRC Immunochemistry Unit at Oxford University from 1967–85 he encouraged, and participated in, a variety of research projects involving the function, structure and organization of the early-acting components of the complement system at both the protein and gene levels. This was especially true with respect to studies on the C1 complex, C2 and C4 components of the classical pathway, factors B and D of the alternative pathway, and the control proteins C3b/C4b inactivator (factor I), factor H, C4b-binding protein (C4BP), properdin (P) and also the C3b/C4b complement receptor 1 (CR1). The work on the complement system proteins developed from studies in the Immunochemistry Unit concerned with defining which portion of the IgG molecule was involved in binding the C1 complex and how this binding brought about the activation of the C1r and C1s proenzymes in the C1 complex. It is hoped that this Essay will illustrate how the elucidation of protein and DNA sequences has helped in the understanding of the activation, function and control of the complement system which is one of the major defence mechanisms in the body.

The complement system consists of at least 20 plasma glycoproteins which can be activated by two distinct routes, the classical and alternative pathways (Fig. 1). One of the glycoproteins, C3, is common to both pathways and this, along with another 12 glycoproteins, constitute the 13 *components* of the classical and alternative pathways. It may be seen (Table 1) that C4 and C3 are by far the most abundant of the complement components, and this is consistent with the central roles they play in the activation of the classical and alternative pathways, respectively. Control of this complex system is mediated partly by the seven *control proteins* found in the plasma. Further regulation of the system, and many of the biological effects of the comple-

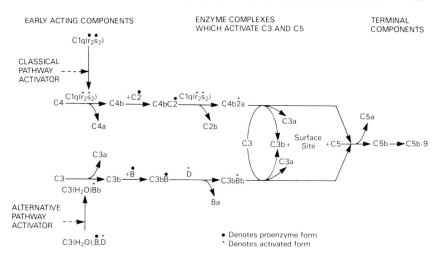

Fig. 1. The major activation steps in the classical and alternative pathways are shown. The system is down-regulated, as outlined in the text, by the following plasma control proteins and membrane-associated regulatory proteins: C1-In (for C1̄r and C1̄s); C4BP and factor I (for C4b); factor H and factor I (for C3b); anaphylatoxin inactivator (for C3a, C4a, C5a); S-protein (for C5b-9 complex); CR1 and factor I (for C3b and C4b); DAF (for C3b and C4b); gp 45-70 and factor I (for C3b). Unlike the other regulators, properdin, which stabilizes C3b,Bb, has an enhancing effect on activation.

ment system (which can involve the induction of inflammatory responses, engulfment, killing and lysis of bacteria and viruses), is mediated through a variety of *membrane-bound receptors*. These can bind to the activated components or to the fragments generated on limited proteolysis of these activated components by the action of the control proteins. Each of these three groups of glycoproteins, namely components, control proteins and membrane-bound receptors, will be described so as to give an overview of the system, with special emphasis being placed on the structural aspects.

Activation of the early-acting components of either the classical or the alternative pathway yields an enzyme complex, a C3 convertase, which can split C3 (Fig. 1). Although the compositions of classical and alternative pathway C3 convertases are different, i.e. C4b,2a and C3b,Bb, they activate C3 in an identical fashion by splitting one peptide bond in the α-chain of C3 to yield C3a and C3b. The C3 convertases can then act as C5 convertases by the association of the C5 with surface-bound C3b, thus allowing the splitting of C5 and the assembly of the C5b-9 lytic complex via the terminal components which are common to both pathways (Fig. 1).

TABLE 1

Plasma proteins involved in activation and control of the complement system

	M_r	Approximate serum concentration $(mg\,l^{-1})$	Number of chains in plasma form prior to activation	Enzymic site in activated form (and natural substrate split)	
Classical pathway components					
C1q	460 000	80	18 (6A + 6B + 6C)	−	
C1r	166 000	50	2 (identical)	+	(C1s)
C1s	166 000	50	2 (identical)	+	(C4, C2)
C4	200 000	600	3 ($\alpha + \beta + \gamma$)	−	
C2	102 000	20	1	+	(C3)
C3	185 000	1300	2 ($\alpha + \beta$)	−	
Alternative pathway components					
Factor D	24 000	1	1	+	(B)
Factor B	92 000	210	1	+	(C3)
C3	185 000	1300	2 ($\alpha + \beta$)		
Terminal components					
C5	191 000	70	2 ($\alpha + \beta$)	−	
C6	120 000	64	1	−	
C7	110 000	56	1	−	
C8	151 000	55	3 ($\alpha + \beta + \gamma$)	−	
C9	71 000	59	1	−	
Control proteins in plasma[a]					
C1-inhibitor	110 000	200	1	−	
C4b-binding protein	550 000	250	7 (identical)	−	
Factor H	150 000	480	1	−	
Factor I	88 000	35	2 ($\alpha + \beta$)	+	(C4b, C3b)
Anaphylatoxin inactivator	310 000	35	6 ($2\alpha + 2\beta + 2\gamma$)	+	(C3a, C4a, C5a)
Properdin	220 000	20	3 or 4 (identical)	−	
S-Protein	83 000	505	1	−	

[a] Membrane-associated regulatory proteins also play a role in control of the complement system as outlined in the text.

II. Activation of the Early-acting Components of the Classical Pathway of Complement Fixation

A. THE C1 COMPLEX AND ACTIVATORS OF THE CLASSICAL PATHWAY

The C1 complex is composed of the subcomponents C1q, C1r and C1s in the ratio 1:2:2. The C1q molecule does not contain enzyme activity and its function in the C1 complex involves the recognition and binding of activators of the classical pathway, thus allowing activation of the C1r and C1s proenzymes to take place.

Electron microscopy studies have revealed that the C1q molecule has an unusual shape, being composed of six peripheral globular "heads", each of which is joined by a collagen-like connecting strand to a fibril-like central portion[1,2] (see Fig. 2). Determination of the amino acid sequences of the polypeptide chains of C1q and how they are disulphide bonded has allowed the construction of a molecular model which is consistent with the electron microscope studies and provides some insight into how C1q may function.[3,4] The C1q molecule is composed of 18 polypeptide chains of three fundamental types (6A, 6B and 6C), each having an M_r of about 25 000. Disulphide bonds between the A- and B-chains and between pairs of C-chains yields nine dimers, i.e. six A–B dimers and three C–C dimers of M_r 53 000 and 48 000 respectively. These nine dimers associate to yield the intact molecule of M_r 460 000 (Fig. 2). Each of the three types of chain present in C1q is 225 amino acid residues long and contains a region of 81 amino acid residues of a collagen-like (Gly—Xaa—Yaa—) repeating sequence starting close to the N-terminal end. The C-terminal portions, of about 136 residues in each chain, are non-collagen like, and structure-prediction studies indicate that they would form predominately β-type entities. These sequence data, together with the electron microscopy measurements and other evidence for the presence of a triple-helical structure, lead to the proposal that three chains (one A-, one B- and one C-chain) form a triple helix through their collagen-like N-terminal regions, and that their non-collagen-like C-terminal portions form one of the globular "heads". However, the collagen-like (Gly—Xaa—Yaa)$_n$ repeating sequences found in the A- and C-chains are interrupted approximately halfway along the length of each collagen region and therefore, in view of the requirement for glycine at every third position to form a true triple helix, a disruption of each triple helix would be predicted at that point (Fig. 2). These observations indicate that one C1q molecule is composed of six triple helices aligned in parallel throughout half their length, which then diverge for the remainder of their length of triple helical structure; each triple helix then merges into one globular "head" region

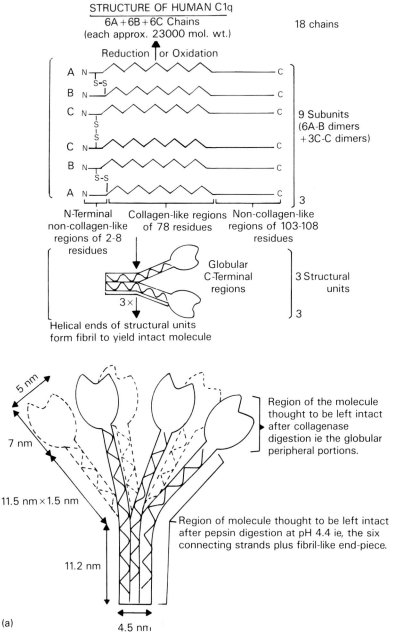

STRUCTURE OF HUMAN C1q

6A + 6B + 6C Chains   18 chains
(each approx. 23000 mol. wt.)

Reduction or Oxidation

9 Subunits
(6A-B dimers
+ 3C-C dimers)

N-Terminal   Collagen-like regions   Non-collagen-like
non-collagen-like   of 78 residues   regions of 103-108
regions of 2-8   residues
residues

Globular
C-Terminal   3 Structural
regions   units

3×

Helical ends of structural units
form fibril to yield intact molecule

5 nm

7 nm

11.5 nm × 1.5 nm

11.2 nm

Region of the molecule
thought to be left intact
after collagenase
digestion ie the globular
peripheral portions.

Region of molecule thought to be left intact
after pepsin digestion at pH 4.4 ie, the six
connecting strands plus fibril-like end-piece.

(a)   4.5 nm

(b)

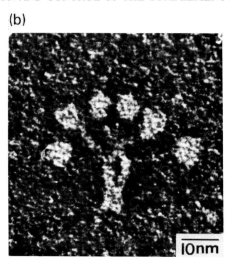

IOnm

Fig. 2. (a) Structure of human C1q. The wavy line represents the proposed triple helix sections, i.e. collagen-like and fibril-like endpiece = 11·5 + 11·2 = 22·7 nm (c.f. value of 23·2 nm predicted from the sequence studies); (—) indicates portions of the molecule pointing towards the reader; (....) indicates portions pointing away from the reader.
  (b) Electron micrograph of a molecule of C1q (from ref. 1) showing the apparent separation of the proposed "structural units". This indicates that the non-covalent bonds which hold the "structural units" together may be located primarily at the N-terminal ends of the chains.

composed of 136 C-terminal amino acids residues of one A-, one B- and one C-chain. This model suggested for C1q (Fig. 2) appears to be close to the conformation it adopts in solution, as judged by neutron diffraction studies,[5] and therefore allows consideration of how it may interact with the $C1r_2C1s_2$ complex and also with activators of the classical pathway.

The $C1r_2C1s_2$ complex is a Ca^{2+} dependent complex of two M_r 83 000 chains of the proenzyme C1r plus two M_r 83 000 chains of the proenzyme C1s. It is easily dissociated by calcium chelators, such as EDTA, to yield a $C1r_2$ dimer plus two C1s chains. The C1s and C1r proenzyme chains have a similar appearance in the electron microscope, both being composed of two globular domains, one slightly larger than the other, with an interdomain linker. The larger domains are considered to contain the catalytic sites in the activated $\overline{C1r}$ or $\overline{C1s}$, while the smaller domains are classed as interaction domains. The complex has therefore the appearance of eight linked globular domains in a slightly inverted S-shape in the electron microscope,[6] the two C1r molecules being located in centre of the chain with their globular catalytic domains in contact. Functional, physical and electron microscopy studies indicate that the $C1r_2C1s_2$ complex interacts with the collagen-like stalks of the C1q molecule.[3,7] On association with C1q, the almost rod-like

$C1r_2C1s_2$ complex may adopt a distorted "figure of 8" structure with the interaction domains of C1r and C1s being located on the outside part of the C1q stalks, while the catalytic domains become located inside the cone defined by the collagen-like stalks of the C1q molecule[7] (Fig. 3). This model of the C1 complex is consistent with the electron microscopy and neutron diffraction data and also appears compatible with what is known about activation, utilization and control of the C1 complex.

The classical pathway may be activated by the interaction of C1 with immune complexes, or aggregates, containing IgG_1, IgG_2, IgG_3 and IgM. However, it is clear that a variety of polyanions (including bacterial lipopolysaccharides, DNA and RNA), small polysaccharides, viral membranes, etc. can activate C1 without the involvement of immunoglobulin.[8] Both activation routes may be of physiological importance, but most work on the mechanism of activation of the C1 complex has employed immune complexes, or aggregates, containing IgG or IgM. In the case of antibody IgG, the evidence available favours a model of C1q–IgG interaction in which

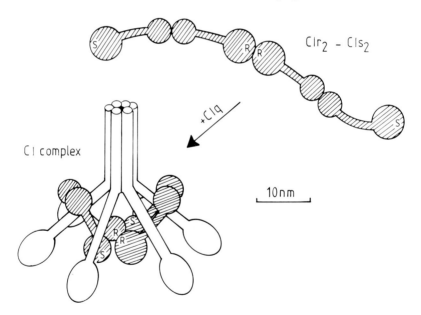

Fig. 3. Models proposed for $C1r_2C1s_2$ and the C1 complex (figure adapted from ref. 7). R and S on the diagram denote the larger, catalytic, domains of C1r and C1s, respectively. The slightly inverted S-shaped $C1r_2C1s_2$ complex consists of a central part made of two monomers of C1r in mutual contact through their globular catalytic domains, while a molecule of C1s is attached to both ends of $C1r_2$. The model of the C1 complex may be formed by placing the rod-like, inverted S-shaped $C1r_2C1s_2$ complex across the C1q arms, then bending each end of the $C1r_2C1s_2$ complex around two opposite C1q arms so that both C1s catalytic domains come into contact with the centrally located catalytic domains of C1r.

the C1q "heads" become firmly attached to multiple Fc regions, presented by the aggregated IgG, without the requirement for a conformational change to have taken place in the Fc regions, i.e. as perhaps might be expected to take place after the interaction of the Fab regions of the IgG antibody with a large-molecular-weight antigen. The primary site on the antibody IgG molecule which interacts with C1q is located in the C_H2 domain of the Fc region and appears to involve an ionic, rather than a hydrophobic, type of binding as judged by experiments using inhibitors of C1q–IgG interaction, specific chemical modification and the labelling of residues in the Fc region which appear important in the binding.[9] The situation regarding the interaction of the C1 complex with IgM is not entirely clear, but the binding of C1q appears to be, through its globular head regions, to the pentameric Fc region of the IgM. In the case of IgM, however, there may be a requirement for a conformational change induced by the exposure of the Fc region by interaction of the Fab arms with antigen.[10]

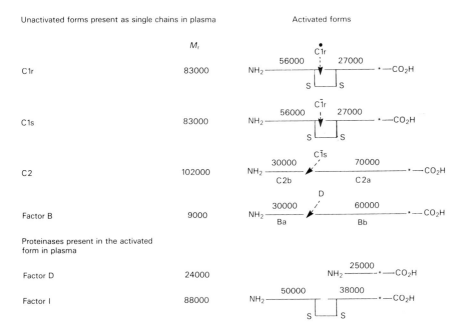

Fig. 4. The serine proteinases involved in the activation and control of the complement system. C1r denotes a single-chain proenzyme form of C1r that is considered to have proteolytic activity. (*) denotes the active-site serine residue which is located about 50 residues from the C-terminus of the catalytic chain of each proteinase.

After C1q has identified and bound to an activator of the classical pathway, activation of the C1 complex can take place. In order for this to occur the $C1r_2C1s_2$ complex has to become associated with C1q. It should be noted that the C1 complex between C1q and the proenzyme forms of $C1r_2C1s_2$ is believed to be weak because it has been estimated that, in normal human serum, up to 30% of the C1q may not be associated with $C1r_2C1s_2$. The inhibitory effect of the C1-inhibitor has of course to be overcome. An efficient activator, by binding to two or more "heads" of C1q, probably causes a conformational change to take place through the collagen-like stalks in C1q, which form the "cage" enclosing the globular catalytic domains of C1r and C1s, thus increasing the affinity of C1q for the $C1r_2C1s_2$ complex. This could allow movement and then autoactivation[4] of the catalytic domains of proenzyme C1r domains which can in turn activate the corresponding catalytic domains of C1s.[7] The active site in the C1 complex (which could be envisaged as being on the outer portion of the $\overline{C1s}$ catalytic domain) splits C4 and then C2. Control of the complex is mediated by the C1-inhibitor which, as a highly elongated molecule, would have access to both the $\overline{C1r}$ and $\overline{C1s}$ active sites, as outlined later (see Section V). The $\overline{C1r}$ and $\overline{C1s}$ form two of the six serine proteinases involved in activation and control of the complement system (Fig. 4).

B. COMPONENTS C4 AND C2 AND FORMATION OF THE C3 CONVERTASE C4b,2a

Component C4 is synthesized as a single, intracellular, precursor protein chain of M_r approximately 200 000.[11,12] Prior to secretion, the pro-C4 is glycosylated, an internal thiolester bond is formed and the molecule is split by processing enzymes to give the major plasma form of three disulphide-linked chains of α (M_r 95 000, βM_r 70 000 and γM_r 33 000 (Fig. 5). Prediction of the entire amino acid sequence for pro-C4 from cDNA studies[11] has shown that the processing events must involve splitting and trimming of short sections containing basic amino acids at the β–α and α–γ junctions (Fig. 5). Activation of C4 by the $\overline{C1s}$ enzyme of the $\overline{C1}$ complex results in the α-chain of C4 being split at one position to yield the C4a peptide ($M_r \sim 9000$) and the large C4b fragment (Fig. 5). The freshly activated C4b can bind covalently, through a reactive carboxyl group of a glutamyl residue in its α'-chain, to a variety of surfaces by the formation of an ester or amide bond.[12–14] This type of binding was first proposed for α_2-macroglobulin[15] and later for freshly activated C3,[12,16,17] as described below. Thus C4, C3 and α_2-macroglobulin all display the following properties: (i) all three can form covalent bonds after activation; they are all inactivated by small amine nucleophiles; (ii) after activation by proteolysis, or inactivation by amines, a

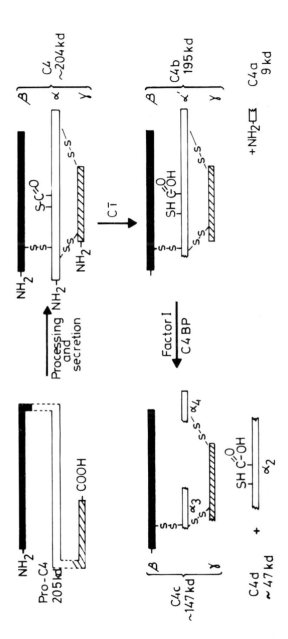

Fig. 5. The biosynthesis, processing and activation of C4. Small areas of the pro-molecule split out by processing enzymes are denoted by a broken line. C̄ls activates C4 by hydrolysis of a single bond in the α-chain, which produces the C4a fragment and simultaneously causes the release of a reactive acyl group and a free sulphydryl group from an intrachain thiolester bond. The acyl group either reacts with water, or can form a covalent bond with hydroxyl or amino groups on a variety of targets such as the cell surface or an antibody molecule. Inactivation of C4b by factor I (and its cofactor C4BP) is caused by the splitting of two bonds in the α'-chain yielding C4c and C4d.

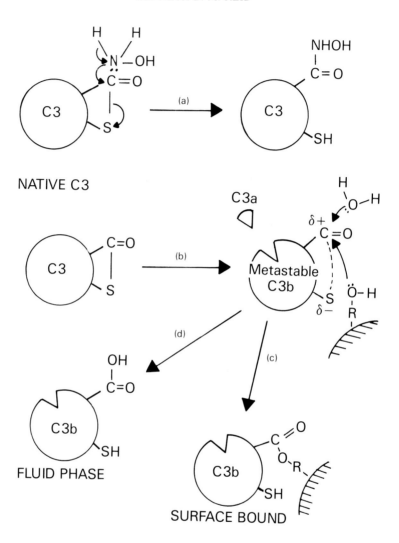

Fig. 6. The covalent binding reaction of C3 (or C4). The thiolester bond which is considered to be present in the α-chain of C3 (and C4) can undergo a variety of reactions : (a) it can be rapidly inactivated by small amines such as NH_2OH (or slow hydrolysis by H_2O can take place giving $C3(H_2O)$); (b) proteolytic activation can occur, by the C3 convertases of either pathway to produce metastable C3b; (c) the reactive acyl group can react with hydroxyl groups (or amino groups) on cell, or particle, surfaces; (d) the reactive acyl group can react with water to form fluid-phase C3b.

free thiol group is found in all three which cannot be detected in the native molecule; (iii) small radiolabelled nucleophiles can be incorporated at a glutamyl residue that is located close to the freshly generated free thiol group. These properties lead to the conclusion that a thiolester bond is present in all three molecules (Fig. 5). The covalent binding of freshly activated, metastable C4b (or C3b) to suitable surfaces, such as membranes or immune aggregates, could proceed in the following manner: the internal thiolester bond which is usually protected in a hydrophobic pocket within the C4, or C3, molecule becomes exposed when proteolytic activation of C4 to C4b, or C3 to C3b, occurs. The freshly activated C4b, or C3b, has the capacity to bind to receptive surfaces through hydroxyl or amino groups by an acyltransfer reaction, thus leading to the formation of ester or amide linkages (Fig. 6). Alternatively the reactive acyl group in C4b or C3b may simply react with water, the most abundant nucleophile, to form fluid-phase C4b or C3b (Fig. 6).

The proenzyme form of component C2 (M_r 102 000) associates, probably through its N-terminal C2b domain, with C4b in an Mg^{2+}-dependent interaction. The C2 is split by $\overline{C1s}$ in the $\overline{C1}$ complex to yield the non-catalytic C2b of M_r 30 000 and the C-terminal catalytic chain C2a of 70 000, the C2b not being required for the C3 convertase activity of the C4b,2a complex.[18] Many of the metastable C4b molecules may become bound near the C1 complex, thus ensuring the formation of the C3 convertase at a location close to the original site of activation. Since the active site of the C3 convertase is located within C2, the attachment of C2 to C4b and the splitting of C2 by $\overline{C1s}$ are obviously important in the generation of convertase activity, and it may be that only C4b bound close to activator surfaces carrying $\overline{C1}$ can form the C3 convertase efficiently. This observation is supported by the fact that although $\overline{C1s}$ is not part of the C3 convertase, it must be present in close proximity to ensure efficient convertase activity,[18] and by the finding that C4 cleavage must precede C2 cleavage if convertase activity is to be generated.

The C4b,2a complex is extremely unstable and therefore its activity is controlled partly by the spontaneous dissociation of C2a from the complex. Control is also mediated by C4b-binding protein (C4BP) and factor I as outlined later.

C. COMPONENT C3

Component C3 is the most abundant protein of the complement system, being present at a concentration of about $1 \cdot 3$ g l^{-1} in plasma. It plays an important central role in complement activation, since it participates in both the classical and alternative pathways (Fig. 1). Many of the biological properties of the complement system, such as opsonizing activity, are mediated principally by the fragments generated during the activation and regulation of C3 by specific complement enzymes, and by other tryptic-like proteinases found in plasma. C3 is synthesized as an intracellular single-chain precursor of M_r 185 000 which contains a thiolester bond similar to that found in C4 and in α_2-macroglobulin.[19] It is then processed to form a two-chain molecule with an α-chain of M_r 115 000 and a β-chain of M_r 70 000. It can be predicted from cDNA studies[20] that in pro-C3 the two chains are arranged in the order β–α and are separated by four contiguous arginine residues which are absent in mature C3 and must therefore have been removed by the proteolytic processing. During activation of C3 by the C3 convertase C4b,2a from the classical pathway, or the C3 convertase C3b,Bb from the alternative pathway, the C3 is cleaved in an identical manner, i.e. the peptide bond between arginine 77 and serine 78 near the N-terminus of the α-chain is hydrolysed. This results in the generation of an M_r 9000 C3a fragment and an M_r 176 000 C3b fragment which contains the remaining C-terminal portion of the α-chain (now called the α'-chain) disulphide-linked to the β-chain. The C3a fragment, like the C4a and C5a fragments, is an anaphylatoxin involved in enhancement of vascular permeability and induction of smooth muscle contraction.[21] The C3b fragment, like C4b, contains an intrachain thiolester bond in its α'-chain and thus in its freshly activated, metastable form can bind covalently to suitable acceptor molecules which are near the site of activation and contain hydroxyl- or amino-groups. The first description of this type of binding was made by Müller-Eberhard et al. in 1966,[22] but a full understanding of the covalent binding of metastable C3b, and C4b, via the thiolester bond in the α'-chain only emerged in the period 1977–80 from binding, labelling and amino acid sequencing studies.[12,16,17,19] These studies showed that on treatment of C3 with small radioactive nucleophiles, such as CH_3NH_2, the radiolabel was incorporated into the penultimate glutamic acid in the sequence —Gly— Cys—Gly—Glu—*Glu*—Asn found in the α-chain, where the residues appearing in italics are considered to form the activatable internal β-Cysteinyl-γ-glutamyl thiolester in C3. On the binding of one mole of nucleophile, one mole of free-SH was exposed which could be labelled and identified by use of radiolabelled iodoacetamide, and thus shown to be in the cysteine residue in the same hexapeptide sequence. In addition, it was known that native C3 contained no free sulphydryl group, while the α'-chain of C3b contains one.

These observations culminated in the chemical synthesis of the hexa-peptide and the demonstration that a thiolester linkage could be formed from this sequence of amino acids. The linkage is sensitive to small nucleo-philes. In native C3 the thiolester is relatively stable, having a half-life of approximately 230 h in isotonic neutral buffer at 37°C,[23] whereas under similar conditions, the half-life of the thiolester in freshly activated meta-stable C3b is about 60 μs.[24] These results show the remarkable degree of reactivity imparted to the thiolester bond, presumably by a conformational change, on the loss of the C3a polypeptide. The high degree of reactivity means that most of the freshly activated C3b will react with water, the most abundantly available small nucleophile. The finding of a thiolester bond in C3 has helped to explain the results of earlier studies which showed that native C3 decayed spontaneously in aqueous solutions and that this decay was accelerated in the presence of chaotropic agents. Since this decay process was accompanied by the appearance of a free sulphydryl which was not present in the native C3, it can be concluded that the decayed C3 is uncleaved C3 in which the thiolester has been hydrolysed, and is usually designated C3i or C3(H_2O). The generation of C3(H_2O) is likely to be an important, if not the major, event in the formation of the initial C3 convertase of the alternative pathway.[23]

III. Activation of the Alternative Pathway

A. FORMATION OF THE INITIAL C3 CONVERTASE

The central enzyme of the alternative pathway is C3b,Bb and it is involved in splitting C3 to produce C3a plus C3b. Thus there is the apparently paradoxical situation where the enzyme complex which generates C3b contains C3b. Lachmann and others[25] proposed that low levels of C3b were continuously generated by enzymic means, leading to a slow "tick-over" of the alternative pathway which could be greatly amplified in the presence of activators of the pathway. This view is essentially correct except that low levels of a "C3b-like" molecule, i.e. C3i, or C3(H_2O), are thought to be first generated by non-enzymic means, due to the spontaneous "decay" of C3 by the slow hydrolysis of the internal thiolester bond in the α-chain of C3.[17,23] Although C3(H_2O) is structurally different from C3b in that it contains an intact α-chain (since the C3a portion is not split off), it behaves functionally in a similar manner to C3b. Both C3(H_2O) and C3b bind to factor B, are regulated by the control proteins factors I and H, and bind to cellular C3b receptors. Therefore the initial convertase of the alternative pathway is probably formed by the interaction of C3(H_2O) with the proenzyme factor B in an Mg^{2+}-dependent manner to form a C3(H_2O),B complex which is then activated by the cleavage of the proenzyme factor B by the enzyme factor D. This would yield a C3 convertase of the form C3(H_2O),Bb which could generate a low level of C3b[17,23] (see Fig. 1).

B. FACTORS D AND B AND FORMATION OF THE C3 CONVERTASE C3b,Bb

Factor D is the smallest complement component, being a single-chain serine protease of M_r 24 000, and it is present in plasma in low concentration (1 mg l^{-1}). The major site of synthesis of factor D is not known but its biosynthesis has been demonstrated in a human monocyte-derived (U937) cell line.[28] It is also not known whether it is synthesized first as a zymogen or in its active form, but in plasma it appears to be present only in its activated form.[29] It is a highly specific serine proteinase, with tryptic-like specificity, splitting a single arg–lys bond in factor B when the latter is bound to C3b. This allows the generation of the C3 convertases C3(H_2O),Bb and C3b,Bb. Factor D is not involved in the enzymic activity of C3b,Bb.

The proenzyme factor B of M_r 90 000 associates, probably through its N-terminal Ba domain, with C3b (or a "C3b-like" form of C3 in the initial events of the alternative pathway activation) in an Mg^{2+}-dependent interaction.[30] Factor B is split by factor D to yield the non-catalytic Ba of M_r 30 000 and the C-terminal catalytic chain of Bb of M_r 60 000. Thus factor D is homologous in function to $\overline{CIs}$ of the classical pathway and factor B and C4b are homologous in function to C2 and C3b of the classical pathway (Fig. 1). This homology is further emphasized by detailed structural studies described later.[31–33]

The C3 convertases of either pathway can "decay" by release of C2a, or Bb, from the C4b or C3b, and this "decay" is accelerated by interaction with control proteins C4BP and factor H, which then act as cofactors in the splitting of C4b, or C3b, by factor I. Various membrane-associated receptor molecules such as CR1, DAF and gp 45–70 can serve as accelerators of the decay of C4b,2a and C3b,Bb. CR1 and gp 45–70 can also serve as cofactors in the splitting of C4b, or C3b, by factor I.

C. ACTIVATORS OF THE ALTERNATIVE PATHWAY

Activators of the alternative pathway include lipopolysaccharides, pure polysaccharides, viruses, bacteria, fungi, tumour cells, red blood cells, parasites, etc.[34,35,36] IgG antibodies can also act as activators after interaction with antigen, F(ab')$_2$ fragments being as efficient as the intact molecule.[3,9] However, it is necessary for the immune aggregates to be in a particulate form rather than in the form of soluble complexes (which can efficiently activate the classical pathway). The precise manner in which activators of the alternative pathway are recognized and function is difficult to postulate given such a wide variety of different types. Absence of sialic acid is considered to be important for activation,[36,37] but in view of the great range of different types of activators it is difficult to identify a principal, common, important structural feature which might be responsible for activation. It is possible that the activators may all offer "protected sites"[34,38] for the deposition of C3b and the formation of a C3b,Bb complex. Any C3b,Bb complex in a "protected site" would then be expected to avoid rapid inactivation by the control protein factor I and its cofactor H. In this manner, a high turnover of C3 can be achieved and the activating particle could become coated with many molecules of C3b derived from freshly activated C3. Thus activation of the alternative pathway can be viewed as a disruption of the inhibitory effects of factors H and I and perhaps enhancement of the stabilizing effect of properdin (P). To emphasize this point, perhaps it should be borne in mind that, since factor D continually circulates in its active form, any recognition in the pathway must reside in the binding of metastable C3b or in the interference with the interaction between C3b and factor H.

As well as being involved in the generation of enzyme complexes having C5 convertase activity, the freshly activated C3b, and C4b, play an important role in the opsonization of micro-organisms or antibody–antigen complexes. The micro-organisms or complexes become coated with large numbers of C3b, or C4b, molecules during complement activation. The opsonized micro-organisms are phagocytosed and killed more efficiently by macrophages, while C3b binding to antibody–antigen complexes results in their solubilization, making them less likely to be deposited in blood vessels and cause tissue injury.

IV. Activation of the Terminal Components C5 to C9

A. FORMATION OF C5 CONVERTASE ACTIVITY

The association of C3b with C5 appears to prepare C5 for attack by the C3/C5 convertases of the classical or alternative pathways.[34,40–42] In both cases it is considered that the binding of one or more molecules of C3b close to the C4b,2a or C3b,Bb enzyme complexes changes the specificity of the complexes from a C3 convertase to a C5 convertase (Fig. 1), and thus the C5 convertases can be considered to have compositions of C4b,2a,C3b and C3b,Bb,C3b. The C2a and Bb portions remain as the active sites in each complex. The required change in specificity is not great as C3 and C5 have very similar structures and in each case a peptide with a C-terminal arginine is cleaved from the α-chain. The cleavage of C5 to C5a and C5b is considered

to be the last proteolytic event in the activation scheme. This scheme
proceeds by a self-assembly mechanism to produce the membrane-attack
complex C5b-9[43,44] (Figs 7 and 8) which can produce channels in a variety of
cells, resulting in cell lysis or killing.

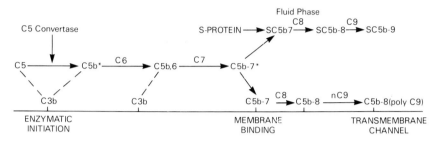

Fig. 7. Scheme of assembly of the membrane attack complex, C5b-9, and its control by
S-protein. The asterisks denote metastable forms of C5b and C5b-7, respectively. This figure is
from ref. 43.

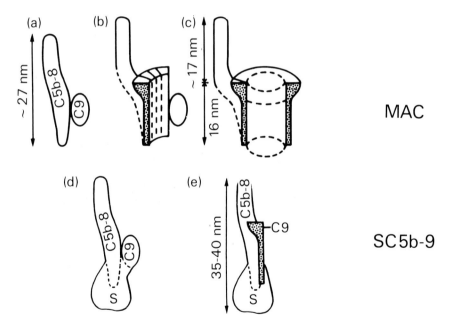

Fig. 8. Hypothetical models for the membrane-attack complex C5b-9 (MAC) and the fluid-
phase complex of S-protein plus C5b-9. Panels (a) to (c): assembly of MAC represented as
C5b-8 complex in association with poly C9. Panels (d) and (e): models for 5C5b-9 containing
C9 in globular (d) or partially unfolded form (e). Taken from ref. 53, with permission.

B. ACTIVATION OF C5 AND ASSEMBLY OF C5b-8

C5, like C3 and C4, is synthesized as a single-chain promolecule which is subsequently processed by cleavage and removal of a short sequence of basic residues, prior to secretion in the form of two disulphide-linked chains, α of M_r 115 000 and β of M_r 75 000. All three proteins (C3, C4 and C5) show regions of strong homology in amino acid sequence, both with each other and with human α_2-macroglobulin, which indicates a common evolutionary origin for these proteins.[45] However, unlike C3, C4 and α_2-macroglobulin, C5 does not contain an internal thiolester bond and consequently does not display covalent binding upon activation, although freshly activated C5 is considered to have a metastable binding site with specificity for C6 due to the transient expression of hydrophobic residues. On activation of C5 by the C5 convertases of either pathway, the peptide C5a of M_r 11 200 and fragment C5b of M_r 180 000 are generated by cleavage of a single arg–leu bond in the α-chain of C5. Like C3a and C4a, C5a is an anaphylatoxin and shows strong leukocyte chemotaxis activity. The freshly activated C5b, containing the metastable binding site for C6, acts as the nucleus of the membrane-attack complex.[46] The C5b, loosely bound to C3b, first binds to C6 forming a C5b-6 complex and then to C7 to form the C5b-7 complex which has a metastable binding site for membrane surfaces[43,44] (Fig. 7). C6 and C7 are both single-chain glycoproteins of M_r about 115 000 and are similar with respect to other physical parameters. Extensive sequence data are not yet available for these proteins, but studies of genetic polymorphisms of C6 and C7 shows that they are closely linked and it has been suggested that they evolved from a common ancestral gene.[47] The C5b-7 complex appears to undergo a hydrophilic–amphiphilic transition upon the binding of C7, thus generating the metastable binding site for membranes and causing the complex to become dissociated from C3b (Fig. 7). The C5b-7 complex is considered to become anchored in the membrane via high-affinity phospholipid binding sites, which are probably located in the C7 molecule. However, the complex does not appear to interfere with membrane function at this stage, as judged by the lack of leakiness of artificial lipid vesicles or of erythrocytes. If the C5b-7 complex fails to bind to a membrane surface its potential cytolytic activity to generate C5b-9 is lost and self-aggregation of the complex takes place in the fluid phase.

C8 of M_r 151 000 contains three chains α, β and γ of M_r 64 000, 64 000 and 22 000, respectively. The α- and γ-chains are disulphide linked and amino acid sequencing has shown that all three chains have different N-terminal sequences. The C8 appears to bind to the C5b-7 complex via its β-chain[48] and this binding probably brings about a conformational change in the C8 molecule, allowing the disulphide-linked α- and γ-chains to penetrate into the hydrophobic core of the lipid bilayer of the membrane to which the C5b-7 complex is attached. Both the α- and β-chains contain higher than normal percentages of hydrophobic amino acids,[49] indicating the probable importance of hydrophobic bonds in the interaction between C5b and the C8 β-chain and between the membrane and the C8 α–γ dimer.[43,44]

C. COMPONENT C9 AND MEMBRANE LESION FORMATION

The role of C5b-8 in the membrane-attack complex is to bind component C9 and also to act as a catalyst in the polymerization of C9 in the formation of membrane lesions.[50] It is estimated that up to 18 C9 molecules can be involved in the formation of the membrane-attack complex which therefore has a composition of $C5b_1$, $C6_1$, $C7_1$, $C8_1$, $C9_n$ (where n can be between 1 and 18). Thus the minimum M_r of the complex is 660 000 and the maximum is 185 000. The C9 molecule, of M_r 71 000, is a single-chain protein whose amino acid sequence has been predicted from the available cDNA sequence.[51] It can be clearly seen that the amino-terminal half of C9 is predominantly hydrophilic in character, while the carboxy-terminal half is more hydrophobic; this is consistent with studies using a lipophilic photo-reactive label which showed that the C-terminal half of C9 is inserted into phospholipid membranes during the formation of membrane lesions.[52] The binding of the first molecule of C9 to the C5b-8 complex is clearly mediated through the C8 component, since this binding is inhibited by anti-C8 antibody and it has been shown that C8 and C9 interact in free solution.[43] The precise mechanism by which C5b-8 causes C9 polymerization is not known but it has to take into account that, in solution, C8 shows only one weak binding site ($K_a \sim 10^7$ M^{-1}) for C9,[50] whereas C8 in the membrane-bound C5b-8 complex can mediate the binding of many C9 molecules and their association is almost irreversible ($K_a \sim 10^{11}$ M^{-1}).[43] Thus the binding of C9 to C5b-8 allows high affinity C9–C9 interaction to take place. The membrane-attack complex $C5b$-$C9_n$ varies in C9, but since this is dependent upon the availability of monomeric C9, if enough C9 is available then polymerization of C9 continues until the typical cylinder-like membrane lesion is completed[53] (Fig. 8). At low levels of C9 relative to C5b-8 (e.g. a ratio of 1 : 1), no typical complement membrane lesions are observed—only a network of large protein aggregates.[50] However, at relatively high ratios of C9 to C5b-8, for example at 6 : 1, discrete ring structures well separated from each other can be seen on the cell membrane surface.[50] Indeed, Dankert and Esser[54] were the first to provide clear evidence that the classical complement "ring-like" lesion may only be incidental to cell lysis and not an obligatory event. They showed that C9 cleaved by thrombin in only one position was as effective as intact C9 in cell lysis, yet produced no "ring-like" structures, only "string-like" structures. Thus it is probable that ring formation is not always a prerequisite for cell lysis and that complexes carrying low numbers of C9 produce smaller hydrophilic membrane channels.

Additional evidence that poly C9 forms the major portion of the typical membrane lesion $C5b-C9_n$ seen, when formed with high concentrations of C9 (Fig. 8), comes from the fact that purified C9 on its own, at 37°C and at neutral pH, in the presence of 50 mM zinc can aggregate and form tubular structures containing between 12 and 18 C9 molecules,[55] although the precise conditions for the induction of self-polymerization of C9 are not entirely clear[56] and it has been suggested that the polymerization process is a side reaction to deactive "active" C9 molecules.[56] However, while polymerizing C9 can attack model lipid vesicles in the absence of C5b-8, it cannot lyse biological membranes unless C5b-8 is present on their surface. This appears consistent with recent results indicating that C5b-8 complex is not only involved with the binding and polymerization of C9, but also participates in the formation of the transmembrane channel structure.[57] Most studies on the formation and function of the transmembrane channel causing cell lysis have been performed using a model target system such as sheep red blood cells. However, little is known about how nucleated cells are killed by complement. Factors such as the capacity of the nucleated cells to respond to complement damage, for example by utilizing repair mechanisms involving lipid synthesis, have to be taken into account.

The remarkable process whereby the five terminal components of the complement system, all soluble plasma glycoproteins, can be transformed into a complex which behaves as if it were composed of integral membrane proteins, may be a common feature in biology. Killer lymphocytes have cytoplasmic granules which contain proteins (channel-forming "perforins"), and these are released upon interaction of the killer cell with target cells. The result is the formation of tubular structures in the target cell similar to those which can be formed with poly C9,[58] although the precise relationship of these polyperforins to C9 is unknown.

Control of the C5b-9 complex is mediated by a control protein,[53] the S-protein, as outlined below.

V. Control Proteins Present in Plasma

A. CONTROL OF ACTIVATED C1 BY C1-INHIBITOR

C1-inhibitor (C1-In) is a single-chain glycoprotein of M_r 104 000 with an unusually high carbohydrate content of 45%. The C1-In is involved in the control of the activation of C1 and also, once the C1 is activated, in the control and disassembly of the active $C\overline{1}r$ and $C\overline{1}s$ molecules in the $C\overline{1}$ complex.[59] The C1-In is considered to interact reversibly with the native C1 complex and prevent spontaneous activation of $C\overline{1}r$ in the complex.[60] Evidence for this is provided by the fact that purified C1 can activate spontaneously but does not do so in the presence of C1-In, for example as in normal human serum. This blocking effect of C1-In can be overcome by efficient activators of the classical pathway such as immune complexes, while activation by non-immune substances such as DNA or heparin is impaired. Once the C1 complex is activated the C1-In rapidly removes the activated $C\overline{1}r$ and $C\overline{1}s$ from the antigen–antibody–$C\overline{1}$ complexes in the form of a $C\overline{1}rC\overline{1}s$-(C1-In)$_2$ complexes.[59] The C1q is left bound to the antigen–antibody complex, probably by means of the "heads", which leaves the collagen-like regions available for interaction with cell receptors or macromolecules such as fibronectin.[3]

B. C4b-BINDING PROTEIN AND FACTOR I
IN REGULATION OF C4b

C4b-binding protein (C4BP) of M_r approximately 550 000 regulates the C3 convertase (C4b,2a) of the classical pathway and is composed of seven identical chains of M_r 70 000. C4BP has a high affinity for C4b and can bind strongly to both fluid-phase and surface-bound C4b.[61] Through this strong association C4BP can efficiently inhibit the formation of the convertase; it can also accelerate the decay of the convertase by displacing C2a from C4b.[61] Once C4BP has bound to C4b it acts as a cofactor by modulating the C4b such that it is readily cleaved by factor I, in two positions, to yield C4c and C4d (Fig. 5). After cleavage by factor I the C4b loses it haemolytic and opsonic activities. Factor I is synthesized as a single-chain precursor of M_r 88 000 and is then processed into two disulphide-linked chains of M_r 50 000 and M_r 38 000, respectively. Although it is a serine protease, which circulates in the plasma, in its active form it is resistant to the classical serine esterase inhibitor diisopropyl-phosphofluoridate. Definitive evidence that it is an enzyme of serine proteinase type was obtained by amino acid sequence studies of the M_r 38 000 chain which was shown to carry the characteristic catalytic site found in all serine esterases.[62] In addition to C4BP, the membrane-associated regulatory proteins complement receptor 1 (CR1), decay accelerating factor (DAF) and glycoprotein 45-70 (gp 45-70) can all serve as accelerators of the decay of C4b,2a, and in the case of CR1 and gp 45-70 can serve as cofactors in the splitting of C4b by factor I.

C. FACTOR H AND FACTOR I IN REGULATION OF C3b

Factor H is a single-chain glycoprotein of M_r 150 000 which binds to C3b and thus regulates many of the functions associated with C3b.[63] It greatly accelerates the decay of the C3b,Bb complex and also the C3b,Bb,P complex (where the enzyme is stabilized by properdin). It is probable that factor H also regulates the C5 convertase since it competes with the binding of C5 to C3b. Factor H, in a similar manner to C4BP in the classical pathway, acts as a cofactor in the cleavage of C3b and also C3 (H_2O) by factor I. Factor I cleaves the α'-chain of C3b in two positions to yield iC3b. Further proteolysis of the iC3b by trypsin-like enzymes yields C3c and C3d, or when the C3bi is bound to CR1, the fragments C3c and C3dg are released by factor I.

D. ANAPHYLATOXIN INACTIVATOR

The structurally and functionally homologous anaphylatoxins C3a, C4a and C5a are released from the α-chains of C3, C4 and C5 upon activation. They are all peptides of 74–78 amino acid residues long which display many inflammatory responses and have been implicated in the regulation of immune responses.[21] Interaction of these anaphylotoxins with cellular receptors brings about vasoconstriction and vascular permeability changes, the spasmogenic activities of the three anaphylatoxins being C5a > 3a > C4a.[21] The C-terminal region of each anaphylatoxin is important for function and, for example, synthetic peptides composed of only the five C-terminal residues of C3a can be shown to have activity. The C-terminal arginine, found in all anaphylatoxins, is essential for activity, and control is therefore efficiently mediated by the anaphylatoxin inactivator (the carboxypeptidase N found in human serum) which selectively removes C-terminal arginine residues from peptides.[64,65]

E. S-PROTEIN IN THE CONTROL OF FORMATION OF THE C5b-9 COMPLEX

The S-protein is a single chain plasma glycoprotein of M_r 80 000 (although variable amounts of a form composed of two disulphide-linked chains of M_r 69 000 and M_r 12 000 is observed). It binds to the metastable C5b-7 complex, thus preventing the complex from binding to the cell surface via its interaction with membrane lipids,[66] and so protectiung bystander cells against lysis by the membrane-attack complex. Up to three molecules of the S-protein are bound per C5b-7 complex and the resulting fluid-phase SC5b-7 complex can then bind C8 and C9 to form SC5b-9, but polymerization of C9 does not take place.[53] The S-protein may also play a role in the blood-clotting system since it is found in a trimolecular complex with thrombin and antithrombin III in serum (but not in plasma).

F. PROPERDIN AND STABILIZATION OF C3b,Bb

Properdin appears to be present in plasma in the form of a mixture of polymers of an M_r 56000 glycoprotein chain. Experiments involving both purified properdin and the assay of properdin activity in whole serum indicate that a mixture of cyclic dimers, trimers, tetramers and higher polymers of the M_r 56000 chain are found, but that the tetramer and trimer appear to be the most common species.[67,68] The properdin monomer appears highly asymmetric when viewed in the electron microscope (26 nm long × 3 nm wide) and it associates in a head-to-tail fashion to form the cyclic polymers.[67] It is unusual in that glutamic acid, glycine and proline account for almost 40% of its amino acid composition. Although over half the sequence has been determined, no molecular model has yet been proposed, but it is clear that there are no collagen-like regions in the molecule.[68] The properdin functions by binding to and stabilizing the C3/C5 convertase of the alternative pathway.

VI. Membrane-associated Regulatory Proteins

A. THE C3b/C4b RECEPTOR (CR1)

CR1 is a single-chain integral membrane glycoprotein of M_r varying between 160000 and 250000 (there are four known allotypes which vary in size by about M_r 30000).[69,71] It is found on red blood cells, lymphocytes and macrophages and it functions by binding C3b, or C4b, in the fluid phase or after they have become covalently bound to particles, cell surfaces or immune complexes.[69,70] It thus acts as a receptor for complexes containing C3b and functions in this manner, for example in neutrophils, by being involved in the release of histamine or, in the case of red blood cell CR1, in the transport of C3b bearing immune complexes to the liver. The purified receptor behaves in a similar manner to both factor H and C4BP in that it increases the decay of C3b,Bb and C4b,2a and it acts as a cofactor in the cleavage of both C3b and C4b by factor I. Recently a plasma form of CR1, which appears identical in structure and function to the membrane form, has been described.[71]

B. DECAY ACCELERATING FACTOR (DAF)

DAF is a single-chain integral membrane protein of M_r 70000 found on red blood cells, platelets and leukocytes.[72] Purified DAF accelerates the decay of both C3b,Bb and C4b,2a, but does not exhibit any cofactor activity in the presence of factor I. The DAF appears to bind only to C3b or C4b molecules on the membrane of the same cell on which the DAF is located, and it may be that its role is to prevent assembly of the convertases rather than to dissociate complexes that have already been formed.[73]

C. GLYCOPROTEINS 45-70 (gp 45-70)

Gp 45-70 is a group of proteins having molecular weights in the range M_r 45000–70000 found in the membranes of leukocytes and platelets and which have ligand binding activity for C3i, C3b, C4b, but not C3d, indicating that the binding region is probably within the C3c portion of C3b.[73] Gp 45-70 may also serve as a cofactor in the cleavage of C3b by factor I and thus may serve a regulatory role in controlling complement activation on cell surfaces, but with preferential activity for C3b and C3b-containing enzymes.

Other membrane-bound receptors in the complement system are the C1q[74] receptor, the CR2 and CR3 receptors (for fragments of C3), the H receptor and C5a receptor.[69,75] These receptors, although of biological importance, are not directly concerned with the main activation pathways and therefore will not be described here.

VII. Molecular Cloning and Genetics

A. INTRODUCTION

The importance of complement as a defence mechanism can be seen from studies of individuals who are genetically deficient in certain components and who, as a consequence, show a marked susceptibility to particular infections. For example, homozygous deficiencies of one of the early-acting components (C1q, C1r, C1s, C2, C4 or C3) is usually associated with the development of immune complex disease.[47] The reason for the increased susceptibility to immune complex disease of these homozygous-deficient individuals may be related to the observation that effective solubilization and clearance of the immune complexes requires the deposition of activated C3 via the classical pathway. It is well known that the bound fragments of activated C3 and C4, i.e. C3b and C4b, can participate in immune adherence and opsonization reactions by binding to specific receptors on certain types of cells, and can thus enhance phagocytosis of foreign particles. Another factor which may be of importance is that, in the case of C2, C4 and factor B, certain polymorphic variants of the genes for these components are associated with disease susceptibility, especially for autoimmune diseases, which is indicative that particular alleles of these genes may influence the effectiveness of the immune response.[76] Deficiencies resulting in C3 dysfunction lead to a great susceptibility to bacterial infections, while deficiencies of C5 to C9 result in a susceptibility to neisserial infection.[47]

In the past four years C1q, C1r, C1s, C2, C3, C4, C5, C9, factor B, C4BP, factor H, C1-In, CR1 and S-protein have been cloned.[77,78] The initial cDNA clones in most cases were obtained from liver cDNA libraries (except the CR1 clone, which was obtained from a tonsil cDNA library) and, in general, the use of mixed-sequence oligonucleotide probes (predicted from available amino acid sequence) has proved the most successful method. The remaining components and receptors seem likely to be cloned in the near future given the general availability of amino acid sequence data (to synthesize oligonucleotide probes) and good polyclonal and monoclonal antibodies (to screen cDNA expression libraries). The cloned DNA will be of value in: (i) establishing the precise molecular defect in deficiency states; (ii) examination of polymorphic variants of a particular component and investigating their possible relationship to disease states; (iii) establishing chromosomal location of various components; (iv) examination of control of level of expression and biosynthesis in different tissues; (v) *in vitro* expression studies by transfection of mammalian cells with cloned complement genes; (vi) examination of the structure and processing of precursor molecules; and (vii) provision of insight into the evolution of families of proteins such as the serine proteinases (of which there are six in the complement system—Fig. 4) by study of intron/exon boundaries. Studies of these types have successfully been initiated in the case of the three components C4, C2 and factor B, which are closely linked and coded for by genes in the major histocompatibility complex on chromosome 6 in man. The study of another gene cluster of functionally related proteins, C4BP, factor H and CR1 on chromosome 1, has also been greatly advanced by the availability of cDNA and genomic clones. A brief description of these two gene clusters and also the cloning of C1q, C3, C5, C9 and S-protein will be given.

B. SUBCOMPONENT C1q

Both cDNA and genomic clones have been isolated from a liver cDNA library for the B-chain of C1q.[79] Analysis of the cDNA and mRNA (from Northern blotting) indicates that there is probably no large pro-form of this

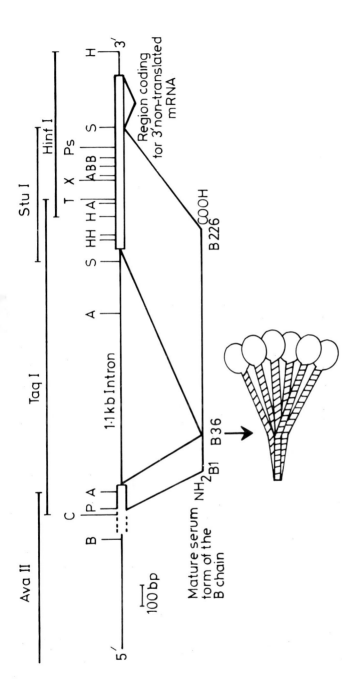

Fig. 9. Restriction map of the C1q B-chain gene. The continuous-line boxed regions show the definitely established exon sequence coding the B-chain mRNA; the broken-line box shows the exon sequence coding for the predominately hydrophobic amino acid sequence. The precise extent of the 5' exon (or exons) has not yet been defined. Restriction sites used in the characterization of the gene are shown: A, *Ava*II; B, *Bst*EII; C, *Cla*1; H, *Hin*f1; P, *Pvu*II; Ps, *Pst*1; S, *Stu*1; T, *Taq*1; X, *Xho*1. The position of the glycine residue at B36 in each of the six B-chains present in one molecule of C1q is indicated by the arrow (↓). This is the position where there are interruptions in the Gly—X—Y repeating amino acid sequences of the A- and C-chains and also where the molecule appears to bend when viewed in the electron microscope.

chain synthesized in the liver, although biosynthetic studies with fibroblast cultures suggested that a type of C1q with chains of $M_r \sim 48\,000$ are synthesized. The C1q gene is about 2·6 kb long and only one intron (of 1·1 kb) has been found in the coding region of the gene: it is located within the codon for glycine 36 (i.e. the middle of the collagen-like region of the C1q B-chain—see Section II.A). This intron is positioned exactly at the point where the triple-helical chains of C1q appear to bend when viewed in the electron microscope (Fig. 9) and it will be of interest to see if the genes for the A- and C-chains follow a similar exon/intron boundary pattern. Unlike the human fibrillar collagens, the triple helical region of the C1q B-chain does not follow the general rule of being encoded by exons of 45, 54, 99, 108 or 162 bp, corresponding exactly to repeats of the Gly—X—Y amino acid sequence repeat.

Two types of genetic deficiency of C1q function, resulting in an inability to deal with immune complexes, have been described: in one there is a non-functional C1q molecule which has antigenic activity; in the other type no C1q antigenic activity can be demonstrated. By the use of the cDNA for the B-chain it can be shown in one family that the second type of deficiency is probably a consequence of an inability to transcribe the B-chain, since this patient's genomic DNA, in the region coding for the globular C-terminal portion, shows a different restriction enzyme pattern on Southern blotting compared to normal individuals.[80] However, further studies utilizing DNA probes for the A- and C-chains and examining the processing events involved in producing the entire C1q molecule are required to provide a good understanding of these deficiency states.

C. COMPONENTS C4, C2 AND FACTOR B

These components are designated the class III complement genes in the major histocompatibility complex and are present, along with the 21-hydroxylase genes, as a tight cluster in the HLA system on chromosome 6 in man (Fig. 10). Class I genes code for the histocompatibility antigens, cell-surface glycoproteins of $M_r \sim 43\,000$ which, in association with β_2-microglobulin, are found at the surfaces of most nucleated cells. Class II genes also code for cell-surface glycoproteins, but show a much more restricted distribution and are composed of two different polypeptide chains of M_r 33 000 and M_r 29 000. The class I and II antigens are highly polymorphic and are involved in the regulation of the immune response. They are related to each other and also to other proteins in the immunoglobulin superfamily.[81,82] However, the complement class III proteins, C4, C2 or factor B, show no similarity in structure to the class I and II gene products.[78] As pointed out

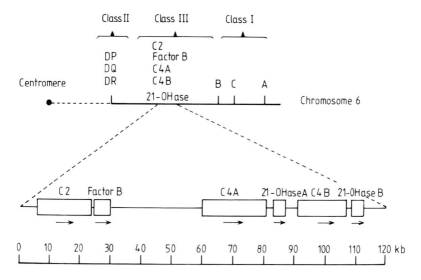

Fig. 10. Map of the major histocompatibility complex in man (HLA) on the short arm of chromosome 6. The class III region is expanded to show alignment of the complement genes C2, factor B, C4A, C4B and also the P_{450} 21-hydroxylase coding sequences, 21-OHase A and 21-OHase B.

earlier (Section II.B), the genes for C2 and factor B code for serine esterase-type proteinases of M_r 108 000 and M_r 92 000, respectively. Component C4 is quite unlike C2 and factor B, but shows structural and functional homology with C3, C5 and the protease inhibitor α_2-macroglobulin, although none of these latter proteins are coded for by genes linked to the major histocompatibility complex.

In one respect, the complement class III genes are similar to the class I and II genes—namely, in the degree of polymorphism which they show. C4 is highly polymorphic, being coded for by at least two separate loci, C4A and C4B, with 35 types having been assigned so far, while factor B has 11 alleles at one locus and C2 has four alleles at one locus. There is a marked association between particular alleles in the HLA-B and HLA-D loci and increased susceptibility to certain diseases, generally autoimmune in character. The possible role of certain types of the complement class III antigens in the susceptibility of an individual to various disorders such as ankylosing spondylitis, myasthenia gravis, systemic lupus erythmatosus and Type 1 diabetes, is currently being further evaluated using the cDNA probes which have recently become available.

(1) C4

There are two types of C4—C4A and C4B—and in most individuals each type is coded by loci approximately 10 kb apart on both chromosomes. Expression of only one locus on one or both chromosomes is observed, as is duplication of loci on either chromosome. After activation by $\overline{C1s}$, the C4A and C4B types show differences in covalent bond formation (via the reactive acyl group released from the thiolester bond) with the amino or hydroxyl groups of small molecules, C4A reacting much more rapidly with amino groups than C4B, and vice versa for hydroxyl groups.[83] An assessment of the different alleles of C4A and C4B has been made possible by the availability of cDNA and genomic clones and DNA sequencing. It has been found that the allelic forms of C4 appear to differ in less than 1% of their amino acid sequences and that these differences (approximately 15 out of 1722 residues) are found almost entirely in within a section of 230 residues in the α'-chain, C-terminal to the thiolester bond[84] (Fig. 5). This section of the α'-chain may be important in determining the reactivity of the C4A and C4B alleles, and hence to the effectiveness of an individual's complement system, due to the central role played by C4 in its ability to interact with a wide variety of proteins, i.e. antigen, antibody, $\overline{C1s}$, C2, C3, C4BP, factor I, and CR1.[76,85] Thus these small differences in amino acid sequence could affect markedly the potential reactivity of the different C4 alleles and have profound biological effects, for example on an individual's ability to deal with immune complexes by the classical pathway.

(2) C2 and Factor B

The C2 and factor B proteases are very similar in gross structure (showing about 35% amino acid sequence homology) and function, which is not too surprising since the genes coding for them are less than 1 kb apart and presumably arose by gene duplication.[31–33,86] The C2 gene is located 5' to the factor B gene (with only 450 bp separating the two genes) and in view of the fact that the concentration of factor B in the blood is about 15 times that of C2, a study of the elements controlling the transcription of the factor B gene will be of interest. The entire 6·5 kb gene structure of human factor B has been determined (Fig. 11) and certain interesting features have emerged from an examination of the intron/exon boundaries of the gene.[33] Examination of the amino acid sequence of the N-terminal Ba fragment, as derived from the cDNA sequence, showed that Ba is composed of three internal homology regions of about 60 amino acids,[32] and the gene structure shows that these regions are precisely encoded by separate exons (Fig. 11). A similar situation is found in C2b, the N-terminal portion of C2.[86] The

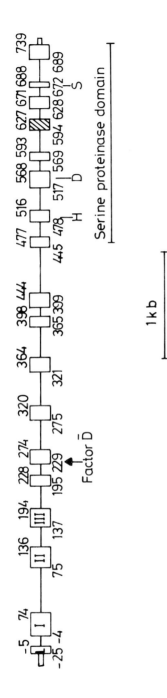

Fig. 11. Structure of the factor B gene. Exons are boxed, and the numbers refer to the amino acids coded for by each exon. L denotes the exon coding for most of the putative leader peptide; I, II and III are the exons coding for the three homologous regions in Ba. The arrow denotes the position in the exon enclosing the factor D cleavage site. H, D and S denote the positions of the codons for the active site residues (histidine, aspartic acid and serine, respectively) distributed in the eight exons coding for the serine protease domain in the C-terminal half of Bb. The shaded exon has no homologous counterpart on comparison with several other serine proteases and may code for a section of the protein important in the activation or specificity of factor B. Taken from ref. 78.

function of these homology regions is not clear but similar regions, as described later, are found in C4BP, factor H and CR1. A common feature of all these proteins, as well as factor B and C2, is their ability to interact with the C3b or C4b fragments (Fig. 1). The C2 gene is about 18 kb long and therefore is much larger than the factor B gene, but this appears to be due to the presence of larger introns rather than a radical difference in intron/exon boundary organization. The factor B and C2 enzymes are unusual types of serine proteinases in that their catalytic chains are over twice the size of the catalytic chains of most other members of the serine esterase family (Fig. 4). The N-terminal half of Bb, the catalytic chain of factor B, is encoded by five exons, and this region shows no homology with other serine proteinases[33] (Fig. 11). The C-terminal half of the catalytic chain is homologous to the catalytic chains of other serine proteinases and each of the functionally important parts of the active site are encoded by separate exons[33] (Fig. 11). Comparison of this region of the factor B gene with the exon organization of other serine proteinases shows a close correlation between them, but also reveals the presence of an extra exon in factor B which has no homologous counterpart in other serine proteinases such as chymotrypsin, elastase, kallikrein and trypsin. It has been speculated that the role of the section of protein coded for by this exon may be to limit the specificity of factor B to the cleavage of only C3 and C5, or it could play a role in the manner in which the proenzyme factor B becomes activated.[33]

D. C4b-BINDING PROTEIN, FACTOR H AND CR1

Linkage analysis of allotypes of human C4BP, factor H and CR1 indicates that they are closely clustered[73,87] and chromosome mapping studies have assigned them to chromosome 1 in man.[88] Thus there is a gene cluster of functionally related regulatory complement proteins in man similar to the C2, factor B, C4 cluster of components, but on a different chromosome. An interesting feature which has emerged from detailed structural studies of C4BP, factor H and CR1 is that, in addition to showing functional homology,[73] they exhibit an unusual type of structural homology that is shared with the components C2 and factor B and two other non-complement proteins, β_2-glycoprotein I and interleukin-2 receptor.[89] All these proteins contain internal repeating units of approximately 60 amino acids, each having a characteristic framework of highly conserved residues consisting of one tryptophane, two proline and four half-cystine residues (Fig. 12; Table 2). Two other positions show conserved glycine residues, while at other positions a bulky hydrophobic amino acid such as tyrosine or phenylalanine is often found.

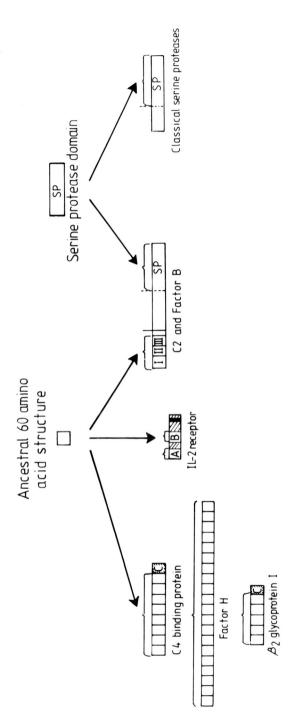

Fig. 12. Scheme showing homology of C2 and factor B to two unrelated protein families, the classical proteases and the 60 amino acid repeat family. The individual repeats are boxed in C4BP, factor H, β_2-glycoprotein 1 and the interleukin-2 receptor. Repeats in C2 and factor B are boxed and numbered I, II and III; repeats in interleukin-2 receptor are labelled A and B. SP, serine protease domain; C, unrelated C-terminal domains in C4BP and β_2-glycoprotein 1. The solid block in the interleukin-2 receptor indicates the transmembrane region.

E. STRUCTURAL EVIDENCE THAT C2, FACTOR B, C4BP, FACTOR H AND CR1 ARE PART OF A SUPERFAMILY OF RELATED PROTEINS

The N-terminal Ba and C2b portions (of factors B and C2) both contain three of the repeating units (described above).[32,33,86] The chains of C4BP[90] and factor H[91] contain eight and 20 repeating units, respectively, and it is thought that CR1 could contain up to 36 of these units (Table 2). The presence of these repeating units thus appears to correlate with functions involving interaction with C3b or C4b. However, since the non-complement proteins β_2-glycoprotein I[93] and interleukin-2 receptor[94] also contain these repeating units (5 and 2 respectively), and it is now known if they show any affinity for C3b or C4b, it may be that the repeating units are a general feature of a superfamily of structurally related proteins. The interleukin-2 receptor is an M_r 55 000 glycoprotein found on the membranes of antigen- or mitogen-stimulated T-cells and thus in the presence of interleukin-2 is involved in the expansion of T-cell populations. The function of β_2-glycoprotein I is not known, but it readily associates with lipoproteins, heparin and binds to platelets.

The repeating units are contiguous and start at the N-terminal end of each processed form of each protein (Table 2, Fig. 12), except in the case of the interleukin-2 receptor where the two repeats in the molecule are separated by a region of 38 amino acids which does not show any homology with the repeating unit.[94] In the case of CR1 the extent of the repeating units throughout the molecule is not known yet, but approximately 36 could be present if, as in factor H, they are found throughout the entire length of the M_r 250 000 form of the CR1 molecule. It has been suggested that duplication or deletion events involving these repeat units may explain the structural polymorphism seen in CR1, i.e. the allotypes of M_r 160 000, 190 000, 220 000 and 250 000. This is an attractive suggestion since analysis of the genes coding for factor B and the interleukin-2 receptor has shown that the repeating homology units in each protein are each coded for by a discrete exon (I, II and III in Ba; A and B in the interleukin-2 receptor, Fig. 12), and initial studies on the C4BP and C2 genes indicate that these genes follow the same rule. It is probable that duplication and switching of exons may have been an important feature of protein evolution.[95] It is notable in the above examples that exon switching occurs between unlinked as well as linked genes which code for proteins having different major biological activities (e.g. the proteinase C2 versus the binding control protein C4BP), although they appear to share some related functions (i.e. binding to C4b or C3b).

Factor H, the chains of C4BP and β_2-glycoprotein I all have an elongated structure and this may be a common feature of all proteins containing large numbers of these repeats.[89]

TABLE 2

Proteins for which there is structural evidence for repeating homology units

Proteins	Structural details	Interacts with C3b or C4b
Complement proteins		
Factor B	3 contiguous units starting at N-terminus of M_r 92 000 chain	C3b
C2	3 contiguous units starting at N-terminus of M_r 108 000 chain	C4b
C4b-binding protein (composed of 7 identical chains)	8 contiguous units starting at N-terminus of M_r 70 000 chain	C4b
Factor H (mouse)	20 contiguous units throughout entire M_r 160 000 chain	C3b
C3b/C4b receptor (CR1)	Number not yet established, but at least 8 units and perhaps up to 36 throughout M_r 160 000–250 000 chain	C3b/C4b
Non-Complement proteins		
β_2 glycoprotein I	5 contiguous units starting at N-terminus of M_r 50 000 chain	Not known
Il-2 receptor	2 units starting at N-terminus of M_r 55 000 chain (the two repeating units are separated by domain of 39 amino acids)	Not known

General consensus sequence of approximately 60 amino acids seen in most repeating homology units:

4	7		30	32	35		46	50	52	57	59
—Cys—Pro————————————			$\frac{Tyr}{Phe}$—Cys—Gly————				Cys—Gly—Trp—			$\frac{Ala}{Pro}$—Cys—	

Certain gaps and deletions have to be made in order to align every repeat in C2, B, C4BP, H, β_2I and the Il-2 receptor in such a way that they conform to the general consensus sequence and numbering shown above. Although only limited sequence data is available for CR1, it is clear that this protein also contains repeating homology units nearly identical to the consensus sequence found in H.

F. COMPONENTS C3, C5, C9 AND THE CONTROL PROTEIN, S-PROTEIN

Mouse C3 was the first complement component for which cDNA clones were obtained. Complete sequences are available now for both human[20] and mouse,[96] there being 79% identity at the nucleotide level and 77% at the amino acid level. The 27 half-cystine residues are identical in both species and the sequence around the reactive thiolester bond near the centre of the α-chain is highly conserved, as might be expected for such a functionally important area of the protein. The size of the C3 gene is estimated to be about 24 kb[97] and the 5' region of this gene, which presumably includes the promotor region, is of some interest since C3 is an acute phase protein, the concentration of which rises rapidly in inflammatory conditions.

The cloning of C5 has allowed the rapid determination of its sequence and comparison with the sequences of C3 and C4 has emphasized the degree of homology between the three proteins and α_2-macroglobulin. This seems to indicate a common evolutionary origin, although C5 lacks the activable internal thiolester present in C3, C4 and α_2-macroglobulin. The cloning of C9 has provided the sequence of this unusual serum protein, which changes from being a soluble protein to an integral membrane protein upon interaction with the C5b-8 complex.[51,98] The sequence has provided some insight as to how this molecule functions (Section IV.C). One unusual finding from the derived amino acid sequence data,[98] which was not discussed earlier, is that the amino-terminal half of C9 contains a region which shows high homology with the 40-residue long, cysteine-rich domains of the low-density lipoprotein receptor. Since C9 polymerization can be inhibited by low-density lipoprotein the "cysteine-domain" in C9 may be of functional importance in the regulation of the membrane-attack complex.

Clones for S-protein have been isolated from a human liver cDNA expression library[99] and this has allowed the derivation of the entire amino acid sequence (459 residues) and leader peptide (19 residues) of this protein, which is involved in the control of formation of the C5b-9 complex on cells. The sequence data show that S-protein is identical with plasma vitronectin, which had been previously characterized as a member of the family of substrate adhesion molecules such as collagen, fibronectin, laminin and thrombospondin.[100] Thus in addition to interacting with the C5b-7 complex and components of the blood-clotting system (as outlined earlier), it is clear that the S-protein/vitronectin can interact with a variety of cell surfaces and subcellular matrices.

VIII. Conclusion

In the past ten years there have been at least five broad approaches to research on the complement system in which structural information has proved valuable in providing a thorough understanding of the functional data. These areas are: (i) the role of the C1q, C1r and C1s subcomponents in the C1 complex; (ii) the study of the covalent binding reactions of C4 and C3; (iii) analysis of the class III genes on chromosome 6 in man; (iv) analysis of the cluster of regulatory genes CR1, factor H and C4BP on chromosome 1 in man; (v) study of formation and control of the C5b-9 membrane-attack complex. Rodney Porter was very much involved in directing research projects concerned with the first four of these areas and his successful contribution to research on complement during the period 1972–81 has been specifically acknowledged in the following awards: the Linacre Lecture (1975), Hopkins Memorial Medal of the Biochemical Society[101] (1977), Croonian Lecture of the Royal Society[102] (1980), and the Copley Medal (1983). He was especially interested in the functional role of C4 and this molecule was central to his research interests from 1979–85. As more data on C4 accumulated, the possible role that allelic forms of C4A and C4B might play in determining the susceptibility of an individual to autoimmune-type disease became a major interest.[76,78,85] His clear and direct approach to research at the structural level would undoubtedly have helped to achieve a rapid understanding of this complex problem.

REFERENCES

1. Knobel, H. R., Villiger, W. & Isliker, H. (1975). Chemical analysis and electron microscopy studies of human C1q prepared by different methods. *Eur. J. Immunol.* **5**, 78–82.
2. Brodsky-Doyle, B., Leonard, K. R. & Reid, K. B. M. (1976). Circular-diachroism and electron-microscopy studies of human subcomponent C1q before and after limited proteolysis by pepsin. *Biochem. J.* **159**, 279–286.
3. Reid, K. B. M. (1983). Proteins involved in the activation and control of the two pathways of human complement. *Biochem. Soc. Trans.* **11**, 1–12.
4. Porter, R. R. & Reid, K. B. M. (1979). Activation of the complement system by antibody-antigen complexes: the classical pathway. *Adv. Protein Chem.* **33**, 1–71.
5. Perkins, S. J. (1985). Molecular modelling of human complement subcomponent C1q and its complex with $C1r_2C1s_2$ derived from neutron-scattering curves and hydrodynamic properties. *Biochem. J.* **228**, 13–26.
6. Tschopp, J., Villiger, W., Fuchs, H., Kilcherr, E. & Engel, J. (1980). Assembly of subcomponents C1r and C1s of the first component of complement. *Proc. Natl. Acad. Sci. USA* **77**, 7014–7018.
7. Colomb, M. G., Arlaud, G. J. & Villiers, C. L. (1984). Activation of C1. *Phil. Trans. R. Soc. Lond. B* **306**, 283–292.
8. Ziccardi, R. J. (1984). The first component of human complement (C1): activation and control. *Springer Semin. Immunopathol.* **7**, 33–50.
9. Burton, D. R. (1985). Immunoglobulin G: Functional Sites. *Mol. Immunol.* **22**, 161–206.
10. Lachmann, P. J. & Hughes-Jones, N. C. (1984). Initiation of complement activation. *Springer Semin. Immunopathol.* **7**, 12–32.
11. Belt, K. T., Carroll, M. C. & Porter, R. R. (1984). The structural basis of the multiple form of human complement component C4. *Cell* **36**, 907–914.
12. Law, S. K., Lichtenberg, N. A. & Levine, R. P. (1979). Evidence for an ester linkage between the labile binding site of C3b and receptive surfaces. *J. Immunol.* **123**, 1388–1394.

13. Campbell, R. D., Dodds, A. W. & Porter, R. R. (1980). The binding of human complement component C4 antibody-antigen aggregates. *Biochem. J.* **189**, 67–80.

14. Janatova, J. & Tack, B. F. (1981). Fourth component of human complement: Studies of an amine-sensitive site comprised of a thiol component. *Biochemistry* **20**, 2394–2402.

15. Howard, J. B., Vermeulen, M., Swenson, R. P. (1980). The temperature-sensitive bond in human α_2-macroglobulin is the alkylamine-reactive site. *J. Biol. Chem.* **255**, 3820–3823.

16. Tack, B. F., Harrison, R. A., Janatova, J., Thomas, M. L. & Prahl, J. W. (1980). Evidence for presence of an internal thiolester bond in the third component of human complement. *Proc. Natl. Acad. Sci. USA* **77**, 5764–5768.

17. Pangburn, M. K. & Müller-Eberhard, H. J. (1980). Relation of putative thiolester bond in C3 to activation of the alternative pathway and the binding of C3b to biological targets of complement. *J. Exp. Med.* **152**, 1102–1114.

18. Kerr, M. A. (1981). The second component of human complement. *Meth. Enzymol.* **80C**, 54–63.

19. Tack, B. F., Janatova, J., Thomas, M. L., Harrison, R. A. & Hammer, C. H. (1981). The third, fourth and fifth components of human complement: isolation and biochemical properties. *Meth. Enzymol.* **80C**, 64–101.

20. De Bruijn, M. H. L. & Fey, G. (1985). Human complement component C3: cDNA coding sequence and derived primary sequence. *Proc. Natl. Acad. Sci. USA* **82**, 708–712.

21. Hugli, T. E. (1981). The structural basis for anaphylatoxin and chemotactic functions of C3a, C4a and C4a. *CRC Crit. Rev. Immunol.* **1**, 321–351.

22. Müller-Eberhard, H. J., Dolmasso, A. P. & Calcott, M. A. (1966). The reaction mechanism of β_{1c}-globulin (C′3) in immune haemolysis. *J. Exp. Med.* **124**, 33–54.

23. Pangburn, M. K., Schreiber, R. D. & Müller-Eberhard, H. J. (1981). Formation of the initial C3 convertase of the alternative complement pathway. Acquisition of C3b-like activities by spontaneous hydrolysis of the putative thiolester in native C3. *J. Exp. Med.* **154**, 856–862.

24. Sim, R. B., Twose, T. M., Paterson, D. S. & Sim, E. (1981). The covalent binding reaction of component C3. *Biochem. J.* **193**, 115–127.

25. Nicol, P. A. E. & Lachmann, P. J. (1973). The alternative pathway of complement activation. The role of C3 and its inactivator (KAF). *Immunology* **24**, 259–275.

26. Lachmann, P. J. & Halwachs, L. (1975). The influence of C3b inactivator (KAF) concentration on the ability of serum to support complement activation. *Clin. Exp. Immunol.* **21**, 109–114.

27. Niemann, M. A., Bhown, A. S., Bennett, J. C. & Volanakis, J. E. (1984). Amino acid sequence of human D of the alternative complement pathway. *Biochemistry* **23**, 2482–2486.

28. Barnum, S. R. & Volanakis, J. E. (1985). In vitro biosynthesis of complement protein D by U937 cells. *J. Immunol.* **134**, 1799–1803.

29. Lesarve, P. & Müller-Eberhard, H. J. (1978). Mechanism of action of factor D of the alternative complement pathway. *J. Exp. Med.* **148**, 1498–1509.

30. Kerr, M. (1981). Human factor B. *Meth. Enzymol.* **80C**, 102–112.

31. Gagnon, J. (1984). Structure and activation of complement components C2 and factor B. *Phil. Trans. R. Soc. Lond. B.* **306**, 301–309.

32. Morley, B. J. & Campbell, R. D. (1984). Internal homologies of the Ba fragment from human complement component Factor B, a class III MHC antigen. *EMBO J.* **3**, 153–357.
33. Bentley, D. R. & Campbell, R. D. (1986). C2 and factor B: Structure and genetics. *Biochem. Soc. Symp. No.* **51**, 7–18.
34. Fearon, D. T. & Austen, K. F. (1980). Current concepts in immunology—the alternative pathway of complement. *New Engl. J. Med.* **303**, 259–263.
35. Reid, K. B. M. & Porter, R. R. (1981). The proteolytic activation systems of complement. *Annu. Rev. Biochem.* **50**, 433–464.
36. Müller-Eberhard, H. J. & Schreiber, R. D. (1980). Molecular biology and chemistry of the alternative pathway of complement. *Adv. Immunol.* **29**, 2–53.
37. Kazatchkine, M. D., Fearon, E. T. & Austen, K. F. (1979). Human alternative pathway: membrane associated sialic acid regulates the competition between β and β1H for cell bound C3b. *J. Immunol.* **122**, 75–81.
38. Fearon, D. I. (1978). Regulation by membrane sialic acid of β1H-dependent decay-dissociation of amplification of C3 convertase of the alternative complement pathway. *Proc. Natl. Acad. Sci. USA* **75**, 1971–1975.
39. Medicus, R. G., Götze, O., Müller-Eberhard, H. J. (1976). Alternative pathway of complement: recruitment of precursor properdin by the labile C3/C5 convertase and the potentiation of the pathway. *J. Exp. Med.* **144**, 1076–1084.
40. DiScipio, R. G. (1981). The binding of human complement proteins C5, factor B, β1H and properdin to complement C3b on zymosan. *Biochem. J.* **199**, 485–496.
41. Isenman, D. E., Podack, E. R. & Cooper, N. R. (1980). The interaction of C5 with C3b in free solution: a sufficient condition for cleavage by fluid phase C3/C5 convertase. *J. Immunol.* **124**, 326–331.
42. Vogt, W., Schmidt, G., van Buthlar, B. & Dieminger, L. (1978). A new function of the activated third component of complement: binding to C5, an essential step for C5 activation. *Immunology* **34**, 29–40.
43. Müller-Eberhard, H. J. (1985). The killer molecule of complement. *J. Invest. Dermatol.* **85**, 47s–52s.
44. Müller-Eberhard, H. J. (1984). The membrane attack complex. *Springer Semin. Immunopathol.* **7**, 93–141.
45. Sottrup-Jensen, L., Stepanik, T. M., Kristensen, T., Lonblad, P. B., Jones, C. M., Wierzbicki, D. M., Magnusson, S., Dimdey, H., Wetsel, R. A., Lundwall, A., Tack, B. F. & Fey, G. H. (1985). Common evolutionary origin of α2-macroglobulin and complement components C3 and C4. *Proc. Natl. Acad. Sci. USA* **82**, 9–13.
46. Lachmann, P. J. & Thompson, R. A. (1970). Reactive lysis: the complement mediated lysis of unsensitized cells II the characterization of activated reactor as C5,6 and the participation of C8 and C9. *J. Exp. Med.* **131**, 643–657.
47. Lachmann, P. J. & Hobart, M. J. (1985). Genetics of complement. *Trends Genetics* **1**, 145–150.
48. Monahan, J. B. & Sodetz, J. M. (1981). Role of the β-subunit in the interaction of the eighth component of human complement with the membrane bound cytolytic complex. *J. Biol. Chem.* **256**, 3258–3262.
49. Steckel, E. W., York, R. G., Monahan, J. B. & Sodetz, J. M. (1980). The eighth component of human complement: purification of physiochemical

characterization of its unusual subunit structure. *J. Biol. Chem.* **255**, 11 997–11 201.

50. Podack, E. R., Tschopp, J., Müller-Eberhard, H. J. (1982). Molecular organization of C9 within the membrane attack complex of complement: induction of circular C9 polymerization by the C5b-8 assembly. *J. Exp. Med.* **156**, 268–282.

51. DiScipio, R. G., Gehring, M. R., Podack, E. R., Kan, C. C., Hugli, T. E. & Fey, G. H. (1984). Nucleotide sequence of human complement component C9. *Proc. Natl. Acad. Sci. USA* **81i**, 7298–7302.

52. Ishida, B., Wisnieski, B. J., Lavine, C. H. & Esser, A. F. (1982). Photolabelling of a hydrophobic domain of the ninth component of human complement. *J. Biol. Chem.* **257**, 10 551–10 553.

53. Podack, E. R., Preissner, K. T. & Müller-Eberhard, H. J. (1984). Inhibition of C9 polymerization within the SC5b-9 complex of complement by S-protein. *Acta Pathol. Microbiol. Scand. [C] (Suppl. 248)* **92**, 89–94.

54. Dankert, J. R. & Esser, A. F. (1985). Proteolytic modification of human complement protein C9: Loss of poly (C9) and circular lesion formation without impairment of function. *Proc. Natl. Acad Sci. USA* **82**, 2128–2132.

55. Tschopp, J., Müller-Eberhard, H. J. & Podack, E. R. (1982). Formation of transmembrane tubules by spontaneous polymerisation of hydrophilic complement protein C9. *Nature (London)* **298**, 534–538.

56. Dankert, J. R., Shiver, J. W. & Esser, A. F. (1985). Ninth Component of Complement: Self-Aggregation and Interaction with Lipids. *Biochemistry* **24**, 2754–2762.

57. Ramm, L. E., Whitlow, M. B. & Mayer, M. M. (1985). The relationship between channel size and the number of C9 molecules in the C5b-9 complex. *J. Immunol.* **134**, 2594–2599.

58. Young, J. D. E., Cohn, Z. A. & Podock, E. R. (1986). The ninth component of complement and the pore-forming protein (perforin 1) from cytotoxic T cells: structural, immunological and functional similarities. *Science* **233**, 184–190.

59. Sim, R. B. & Reboul, A. (1981). Preparation and properties of human C̄1 inhibitor. *Meth. Enzymol.* **80C**, 43–54.

60. Ziccardi, R. J. (1985). Demonstration of the interaction of native C1 with monomeric immunoglobulins and C1 inhibitor. *J. Immunol.* **134**, 2559–2563.

61. Nussenzweig, V. & Melton, R. (1981). Human C4-binding protein (C4-bp). *Meth. Enzymol.* **80C**, 124–133.

62. Yuan, J.-M., Hsiung, L.-M. & Gagnon, J. (1986). Human complement factor I: CNBr cleavage of the light chain and alignment of the fragments. *Biochem. J.* **233**, 339–345.

63. Sim, R. B., Malhotra, V., Ripoche, J., Day, A. J., Micklem, K. J. & Sim, E. (1986). Complement Receptors and Related Complement Control Proteins. *Biochem. Soc. Symp.* **51**, 83–96.

64. Bokish, V. A. & Müller-Eberhard, H. J. (1970). Anaphylatoxin inactivator of human plasma: its isolation and characterization as carboxypeptidase. *J. Clin. Invest.* **49**, 2427–2434.

65. Plummer, H. R. & Hurwitz, N. Y. (1978). Human plasma carboxypeptidase. *J. Biol. Chem.* **253**, 3907–3916.

66. Podack, E. R., Kolb, W. P. & Müller-Eberhard, H. J. (1978). The C5b-9 complex: formation, isolation and inhibition of its activity by lipoprotein and the S-protein of human serum. *J. Immunol.* **120**, 1841–1848.

67. Smith, C. A., Pangburn, M. K., Vogel, C. W. & Müller-Eberhard, H. J. (1984). Molecular architecture of human properdin, a positive regulator of the alternative pathway of complement. *J. Biol. Chem.* **259**, 4582–4588.
68. Reid, K. B. M. (1981). Preparation of human properdin. *Meth. Enzymol.* **80C**, 143–150.
69. Fearon, D. T. & Wong, W. W. (1983). Complement ligand–receptor interactions that mediate biological responses. *Annu. Rev. Immunol.* **1**, 243–271.
70. Dykman, T. R., Hatch, J. A., Aqua, M. & Atkinson, J. P. (1985). Polymorphism of the C3b/C4b receptor (CR1): characterization of a fourth allele. *J. Immunol.* **134**, 1787–1789.
71. Yoon, S. H. & Fearon, D. T. (1985). Characterization of a soluble form of the C3b/C4b receptor (CR1) in human plasma. *J. Immunol.* **134**, 3332–3338.
72. Cole, J. L., Housley, Jr., G. A., Dykman, T. R., MacDermott, R. P. & Atkinson, J. P. (1985). Identification of an additional class of C3-binding membrane proteins of human peripheral blood leukocytes and cell lines. *Proc. Natl. Acad. Sci. USA* **82**, 859–863.
73. Holers, V. M., Cole, J. L., Lublin, D. M., Seya, T. & Atkinson, J. P. (1985). Human C3b and C4b regulatory proteins: a new multi-gene family. *Immunol. Today* **6**, 188–192.
74. Ghebrehiwet, B., Silvestri, L. & McDevitt, C. (1984). Identification of the Raji cell membrane-drived C1q inhibitor as a receptor for human C1q. Purification and immunochemical characterization. *J. Exp. Med.* **160**, 1375–1389.
75. Dierich, M. P. & Schulz, T. (1983). The nature and function of complement receptors. *Progr. Immunol.* **V**, 407–418.
76. Porter, R. R. (1983). Complement polymorphism, the major histocompatibility complex and associated diseases: A speculation. *Mol. Biol. Med.* **1**, 161–168.
77. Reid, K. B. M. (1985). Application of molecular cloning to studies on the complement system. *Immunology* **55**, 185–196.
78. Campbell, R. D., Carroll, M. C. & Porter, R. R. (1986). The molecular genetics of components of complement. *Adv. Immunol.* **38**, 203–244.
79. Reid, K. B. M. (1985). Molecular cloning and characterization of the complementary DNA and the gene coding for the B-chain of subcomponent C1q of the human complement system. *Biochem. J.* **231**, 729–735.
80. McAdam, R., Tenner, A., Spath, P., Chung, P. & Reid, K. (1985). An abnormal *Taq*1 restriction fragment in a C1q-deficient patient is caused by the absence of a restriction site in the B chain coding region. *Complement* **2**, 52 (Abstr.).
81. Steinmetz, M. & Hood, L. (1983). Genes of the major histocompatibility complex in mouse and man. *Science* **222**, 727–735.
82. Kaufman, J. F., Auffray, C., Korman, A. J., Shackelford, D. A. & Strominger, J. (1984). The Class II molecules of the human and murine histocompatibility complex. *Cell* **36**, 1–10.
83. Law, S. K. A., Dodds, A. W. & Porter, R. R. (1984). A comparison of the properties of two classes, C4A & C4B, of the human complement component C4. *EMBO J.* **3**, 1819–1823.
84. Belt, K. T., Yu, C. Y., Carroll, M. C. & Porter, R. R. (1985). Polymorphism of human component C4. *Immunogenetics* **21**, 173–180.
85. Porter, R. R. (1984). The complement components of the major histocompatibility locus. *CRC Crit. Rev. Biochem.* **16**, 1–19.

86. Bentley, D. R. (1986). Primary structure of human complement component C2: homology to two unrelated protein. families. *Biochem. J.* (in press).
87. De Cordoba, S. R., Lublin, D. M., Rubinstein, P. & Atkinson, J. P. (1985). Human genes for three complement components that regulate the activation of C3 are tightly linked. *J. Exp. Med.* **161**, 1189–1195.
88. Klickstein, L. B., Wong, W. W., Smith, J. A., Morton, C., Fearon, D. T. & Weis, J. H. (1985). Identification of long homologous repeats in human CR1. *Complement* **2**, 44–45.
89. Reid, K. B. M., Bentley, D. R., Campbell, R. D., Chung, L. P., Sim, R. B., Kristensen, T. & Tack, B. F. (1986). Complement system proteins which interact with C3b, or C4b: evidence that they are part of a superfamily of structurally related proteins. *Immunol. Today* **7**, 230–234.
90. Chung, L. P., Bentley, D. R. & Reid, K. B. M. (1985). Molecular cloning and characterisation of the cDNA coding for C4b-binding protein, regulatory protein of the classical pathway of the human complement system. *Biochem. J.* **230**, 133–141.
91. Kristensen, T. & Tack, B. F. (1986). Molecular cloning of mouse factor H. *Proc. Natl. Acad. Sci. USA* (in press).
92. Wong, W. W., Klickstein, L. B., Smith, J. A., Weis, J. H. & Fearon, D. T. (1985). *Proc. Natl. Acad. Sci. USA* **82**, 7711–7715.
93. Lozier, J., Takahasi, N. & Putnam, F. W. (1984). Complete amino acid sequence of human plasma β_2-glycoprotein I. *Proc. Natl. Acad. Sci. USA* **81**, 3640–3644.
94. Leonard, W. J., Depper, J. M., Kanehisa, M., Krönke, M., Peffer, N. J., Svetlik, P. B., Sullivan, M. & Greene, W. C. (1985). Structure of the human interleukin-2 receptor gene. *Science* **230**, 633–639.
95. Gilbert, W. (1985). Genes-in-pieces, revisited. *Science* **228**, 823–824.
96. Wetsel, R. A., Lundwall, A., Davidson, F., Gibson, T., Tack, B. F. & Fey, G. H. (1984). Structure of murine complement component C3 nucleotide sequence of cloned complementary DNA coding for the α-chain. *J. Biol. Chem.* **259**, 13857–13865.
97. Wiebauer, K., Domday, H., Digglemann, H. & Fey, G. H. (1982). Isolation and analysis of genomic clones encoding the third component of mouse complement. *Proc. Natl. Acad. Sci. USA* **79**, 7077–7081.
98. Stanley, K. K., Kocher, H.-P., Luzio, J. P., Jackson, P. & Tschopp, J. (1985). The sequence and topology of human complement component C9. *EMBO J.* **4**, 375–382.
99. Jenne, D. & Stanley, K. K. (1985). Molecular cloning of S-protein, a link between complement, coagulation and cell-substrate adhesion. *EMBO J.* **4**, 3153–3157.
100. Edelman, G. M. (1985). Cell adhesion and the molecular processes of morphogenesis. *Annu. Rev. Biochem.* **54**, 135–168.
101. Porter, R. R. (1977). The biochemistry of complement—The Eleventh Hopkins Memorial Lecture. *Biochem. Soc. Trans.* **5**, 1659–1674.
102. Porter, R. R. (1980). The complex proteases of the complement system. The Croonian Lecture. *Proc. R. Soc. Lond. B.* **210**, 477–498.

Processing and Metabolism of Neuropeptides

ANTHONY J. TURNER

MRC Membrane Peptidase Research Group, Department of Biochemistry,
University of Leeds, Leeds LS2 9JT, England

I.	Introduction	70
	A. What are the Neuropeptides?	70
	B. Identification and Localization	72
II.	Selected Neuropeptides	74
	A. Opioid Peptides	74
	B. Substance P and the Tachykinins	79
	C. Calcitonin and CGRP	82
	D. Atrial Natriuretic Factors (atriopeptins)	83
III.	Processing of Neuropeptide Precursors	84
	A. Organization of Neuropeptide Biosynthesis	84
	B. Endopeptidase Cleavage	86
	C. Carboxypeptidase B-like Activity ("Enkephalin Convertase")	88
	D. C-Terminal Amidation	90
	E. Other Modification Processes	90
	F. The Processing of Neuropeptide Precursors is Tissue-specific	91
IV.	Inactivation of Neuropeptides	94
	A. Modes of Neurotransmitter Inactivation	94
	B. Enkephalin Metabolism	94
	C. Microvillar Peptidases as a Model System	95
	D. Synaptic "Enkephalinase" is Identical with Microvillar Endopeptidase-24.11	97
	E. Aminopeptidases can Inactivate Enkephalins and other Susceptible Neuropeptides	101
	F. Angiotensin Converting Enzyme (Peptidyl Dipeptidase A) is also common to Renal and Synaptic Membranes	101
	G. Summary	103
V.	Design of Peptidase Inhibitors	104
	A. Inhibition of Neurotransmitter Metabolism	104
	B. Carboxypeptidase A and Angiotensin Converting Enzyme	105
	C. Thermolysin and "Enkephalinase"	106
VI.	Likely Future Developments	109
	Acknowledgements	110
	References	110

ESSAYS IN BIOCHEMISTRY Vol. 22
ISBN 0 12 158122 5

Abbreviations

AcCh	acetylcholine	GABA	γ-aminobutyric acid
ACE	angiotensin converting enzyme	GEMSA	guanidinoethylmercaptosuccinic
ACTH	adrenocorticotropic hormone		acid
ANF	atrial natriuretic factor	LPH	lipotropin
CCK	cholecystokinin	MSH	melanocyte stimulating hormone
CGRP	calcitonin gene-related peptide	POMC	pro-opiomelanocortin
CLIP	corticotropin-like intermediate	SER	smooth endoplasmic reticulum
	peptide	SSA	succinic semialdehyde
CNS	central nervous system	TRH	thyrotropin releasing hormone
CP-A	carboxypeptidase A	VIP	vasoactive intestinal peptide
CRF	corticotropin releasing factor		

I. Introduction

A. WHAT ARE THE NEUROPEPTIDES?

The transfer of signals between neurons or from a neuron to its target cell is almost always mediated by chemical neurotransmitters released at nerve synapses. After depolarization, the neurotransmitter molecule diffuses across the synaptic cleft and binds to specific receptors located on the plasma membrane of the post-synaptic cell. At this site, the transmitter can cause changes in membrane permeability leading to either an excitatory or inhibitory post-synaptic potential. For efficient neuronal communication one might envisage the need for two, or at most a few, transmitters which may be either excitatory or inhibitory in nature. Until recently, this simplistic view appeared to be borne out since only a handful of transmitters had been identified since the discovery of acetylcholine in the 1920s. Indeed, the mammalian cerebellum may function almost exclusively using glutamate and γ-aminobutyrate (GABA) as excitatory and inhibitory signals, respectively.

In addition to acetylcholine, the so-called "classical neurotransmitters" comprise the biogenic monoamines (principally noradrenaline, dopamine, serotonin), as well as certain amino acids (glycine, glutamate, aspartate, γ-aminobutyrate). The discovery during the last decade of more than forty functional peptides in the central and peripheral nervous systems of many species (Table 1) has hence transformed our view of the nature of neurotransmission and of its regulation.[1,2] These discoveries have largely arisen as a result of recent advances in techniques for protein isolation and microsequencing as well as from the application of molecular biological techniques to the nervous system. More often than not, the discovery of the peptides has preceded an understanding of their physiological functions. In many cases, it is unclear whether these peptides have a direct action on the neighbouring post-synaptic membrane or have a more diffuse effect on target neurons,

TABLE 1

Examples of mammalian brain peptides

Hypothalamic hormones	*Gastro-intestinal peptides*
Corticotropin releasing hormone	Substance P
Growth hormone releasing hormone	[Met]-enkephalin
Luteinizing hormone releasing	Vasoactive intestinal polypeptide (VIP)
hormone	Cholecystokinin (CCK)
Somatostatin	Pancreatic polypeptide (PP)
Thyrotropin releasing hormone (TRH)	Gastrin
	Motilin
Pituitary hormones	Secretin
Oxytocin	
Vasopressin	*Other peptides*
Corticotropin (ACTH)	Atrial natriuretic factor
β-Endorphin	Bradykinin
Melanocyte stimulating hormones	Bombesin
Luteinizing hormone	Calcitonin gene-related peptide
Growth hormone	Dynorphin
Prolactin	Neuropeptide Y
	Neuropeptide YY
Circulating hormones	Neurotensin
Angiotensin	Neurokinin A (Substance K)
Bradykinin	Neurokinin B
Calcitonin	Diazepam binding inhibitor
Glucagon	δ-Sleep inducing peptide
Insulin	FMRFamide

Adapted from ref. 1. The evidence for some of these peptides in the brain relies on isolation and sequencing. For others, the evidence is merely immunochemical.

thereby modifying the action of one or more primary transmitters. Where the precise action of a peptide or other mediator is unclear, the term "neuromodulator" rather than neurotransmitter has gained currency. Defining the functions of the large array of identified neuropeptides will clearly occupy physiologists and pharmacologists well into the next century. The possibility that aberrations of neuropeptide biochemistry might be implicated in neurological dysfunction and mental disease will also, no doubt, receive much attention.

The application of recombinant DNA technology has revealed that the neuropeptides often occur as multiple products from a single gene, or they may be members of closely related gene families. For example, the morphine-like opioid peptides comprising the enkephalins, endorphins and dynorphins all share the common N-terminal sequence Tyr—Gly—Gly—Phe—. In turn, the neuropeptides interact with one or more classes of post-synaptic receptors which may respond with changes in cyclic nucleotide concentrations, phosphoinositide turnover and associated ion fluxes. Although termed neuropeptides, a number of the compounds listed in Table 1 were originally identified as neurosecretory or endocrine hormones. For example, the pituitary peptides vasopressin and oxytocin are secreted from nerve terminals in the pituitary into the blood where they act as circulating hormones (see e.g. ref. 3). However, these peptides have subsequently been demonstrated in the central nervous system where they may function as neurotransmitters.[4] Likewise, the hormones of the hypothalamus such as thyrotropin releasing hormone (TRH), corticotropin releasing factor (CRF) and somatostatin have been demonstrated in brain regions outside the hypothalamus. The enkephalins, originally isolated from the brain,[5,6] are abundant in the adrenal medulla[7] where they are released along with adrenaline and also serve an endocrine role. A number of other peptides, for example gastrin, cholecystokinin (CCK), substance P and vasoactive intestinal polypeptide (VIP), were originally identified in secretory elements of the gut before their immunochemical identification in the CNS.[2] Likewise, the atriopeptins (atrial natriuretic factors) were identified as cardiac hormones[8,9] before their presence was shown in the brain.[10] In contrast, neurotensin was only identified in the gastro-intestinal tract following its characterization from the brain.[11] More recently, new brain peptides have been identified as a result of nucleotide sequence analysis of peptide precursor genes. Apart from neurotransmitter and endocrine roles for identified peptides, some appear to have additional physiological functions, including potent mitogenic actions on tissue[12] and the ability to modulate the responsiveness of the immune system.[13] Thus, target tissues for peptides are more widespread than originally suspected and while there may still be an organ-specific major site of production, the concept that these sites are unique is now untenable. Perhaps the time has come for more general terms such as "regulatory peptide" or "neuroendocrine peptide" to supersede "neuropeptide" or "peptide hormone". Although such peptides represent a diverse group of compounds, it is clear that there exist common biochemical mechanisms for their processing from precursor polypeptides and their inactivation at the cell surface. These topics are the subject of the present essay. Other aspects of the biochemistry of peptide-secreting neurons have recently been reviewed elsewhere.[14]

B. IDENTIFICATION AND LOCALIZATION

Initially, peptides were isolated from the CNS by classical methods of preparative protein chemistry. Recent advances in analytical methods and the use of high performance liquid chromatography have enabled such procedures to be scaled down considerably. In addition, recombinant DNA technology has now proved a valuable adjunct for identifying peptide precursor mRNAs and the genes that encode them. Such an approach has led to the discovery of novel peptides including calcitonin gene-related peptide (CGRP) in the gene coding for the peptide hormone calcitonin.[15] Since many biologically active peptides have amidated C-termini, Tatemoto[16] devised a novel chemical assay for detecting such peptides and thereby isolated various amidated peptides including neuropeptide Y, peptide YY, peptide HI and galanin, all of which are suggested to function as neuronal peptides.

Once a peptide has been isolated, monoclonal or polyclonal antibodies can be raised to the intact peptide, or specific fragments of it, and these can be used to localize the peptide in the CNS and elsewhere. This technique has been extended by Sutcliffe & Milner[17] to the localization of predicted neuropeptides whose sequences have been deduced from the corresponding cDNA sequences of brain-specific mRNAs. The discrete localization of each of the neuropeptides in the CNS has best been revealed by immunohisto-chemistry. This technique has allowed the demonstration of discrete popu-lations of peptide-containing neurons organized into specific pathways.[2,18] However, caution must be exercised in such localization studies in view of the lack of absolute specificity of antisera. For example, CCK was originally detected in the brain using antibodies not to CCK but to the structurally related peptide gastrin.[19] Pancreatic polypeptide immunoreactivity in brain is apparently due to neuropeptide Y.[20] Such mis-identification can be all too common if specificity of antisera is not carefully checked, particularly among families of closely related peptides. Recently, the development of *in situ* hybridization histochemistry[21] has allowed the use of cDNA probes to visualize neuropeptide mRNA, thereby locating the anatomical sites of neuropeptide gene expression. Such techniques provided the first demon-stration that pro-opiomelanocortin (POMC) mRNA was synthesized directly in certain neurons in the CNS and that the intact peptide was not translocated from its site of synthesis in the pituitary.[22] In neurons, the neuropeptides are found to be particularly concentrated in the nerve terminal region. Even so, the concentrations of neuropeptides in the CNS appear to be several orders of magnitude lower (10^{-12}–10^{-15} mol per mg protein) than the "classical" neurotransmitters such as acetylcholine and the monoamines.[23]

A surprising outcome of neuropeptide localization studies has been the realization that neuropeptides can often coexist in neurons with other neurotransmitters where each component may modulate the action of the other.[24] Such coexistence was first demonstrated in invertebrate systems, but many examples have now been reported in mammalian systems (see e.g.

TABLE 2

Some examples of the coexistence of peptides with classical transmitters

Peptide	Classical transmitter	Location
Cholecystokinin	GABA	Cerebral cortex
Enkephalin	Dopamine	Carotid body
Enkephalin	Noradrenaline	Adrenal medulla
		Rat superior cervical ganglion
Neuropeptide Y	GABA	Cerebral cortex
Somatostatin	GABA	Cerebral cortex
Somatostatin	Noradrenaline	Cat mesenteric ganglia
		Adrenal medulla
Substance P	Serotonin	Rat medulla oblongata
		and Raphe nucleus
TRH	Serotonin	Rat medulla oblongata
VIP	Acetylcholine	Autonomic ganglia

Many other combinations of coexistence have been reported. See for example refs 23, 26, 28.

ref. 25). For example in the cerebral cortex, cells that stain immunochemically for glutamate decarboxylase and that therefore synthesize GABA, can be shown to contain somatostatin, neuropeptide Y and CCK.[26] In the case of CCK and somatostatin, they are known to be released from axon terminals in a calcium-dependent manner and they can act as excitatory agents in the mammalian cerebral cortex.[27] The significance of such coexistence is unclear but Hendry et al.[26] have speculated that in neurons where GABA acts as a classical inhibitory transmitter, the peptides released either simultaneously with GABA, or at different times, could exert a modulatory action on the post-synaptic neuron, for example by action on a second messenger system. A number of examples of coexistence of classical transmitters with peptides is provided in Table 2 and ref. 28.

II. Selected Neuropeptides

A. OPIOID PEPTIDES

The family of peptides that have generated the most intense research interest over the last decade are undoubtedly those that exhibit morphine-like activity and are now generically referred to as opioid peptides. This particular area of peptide biochemistry has been reviewed extensively during the last few years (see e.g. refs 29, 30) and the background will only be considered briefly here. Although morphine (Fig. 1) has been known since at least classical Greek times as a drug to elicit euphoria and deaden pain, its mechanism of action and the hunt for specific receptors for the drug were only of incidental interest to biochemists until the mid-1970s. At that time, Kosterlitz & Hughes,[31] in Aberdeen, isolated two naturally occurring pentapeptides from the brain (Fig. 1) that appeared to be the endogenous ligands for the opiate (morphine) receptors. These peptides were named [Leu]- and [Met]-enkephalin. Shortly afterwards the [Met]-enkephalin sequence was detected in a much larger pituitary protein, β-lipotropin, as well as in a fragment of the latter, β-endorphin (endorphin = "*endo*genous *morphin*e"), which also showed potent opiate activity.[32,33] Because of the structural similarities of these peptides, it was naturally assumed that β-endorphin was the precursor of [Met]-enkephalin, which was derived from it by proteolytic processing. However, no precursor–product relationship could be demonstrated between β-endorphin and [Met]-enkephalin and, in any case, β-endorphin did not contain the sequence of [Leu]-enkephalin. Clearly, the enkephalins were the products of different genes

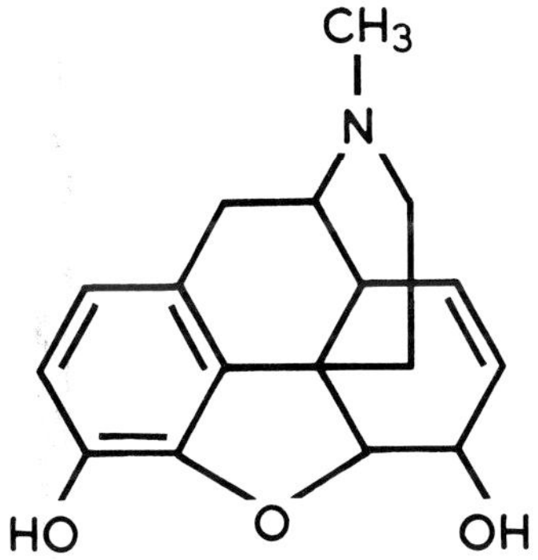

Morphine

$$Tyr^1\text{-}Gly^2\text{-}Gly^3\text{-}Phe^4\text{-}Met^5$$
$$Tyr^1\text{-}Gly^2\text{-}Gly^3\text{-}Phe^4\text{-}Leu^5$$

[Met^5]-and [Leu^5]-Enkephalin

Fig. 1. Structures of morphine and the two prototypes of the opioid family of peptides, [Met]- and [Leu]-enkephalin.

from β-endorphin. Other distinct opioid peptides, such as dynorphin, and α- and β-neoendorphin, were discovered about this time and the relationship between these various peptides has been clarified by recombinant DNA work.

β-Endorphin is synthesized from a much larger (M_r 31 000) precursor protein pro-opiomelanocortin (POMC) that also contains within it the sequences of adrenocorticotropic hormone (ACTH) and several forms of melanocyte stimulating hormone (MSH) (see e.g. ref. 34). Such precursor proteins containing several distinct biologically active peptides have been referred to as polyproteins and the regulation of polyprotein gene expression has been reviewed elsewhere.[34] The origin of the enkephalins was initially elucidated by studying the biosynthesis of the peptides in adrenal medulla, a tissue which turned out to be particularly rich in enkephalins, where they are present at concentrations approaching 20 nmol per mg of tissue. In the adrenal medulla these peptides appear to have a physiological role in regulating catecholamine secretion during times of stress. The opioid peptides can now be grouped into three distinct precursors encoded by three separate genes: POMC, containing the β-endorphin sequence, proenkephalin A and proenkephalin B (or prodynorphin). It is likely that proenkephalin A and B genes have been generated from a common ancestor by gene duplication.[35] The primary sequence of proenkephalin A was deduced from the sequence of the cloned cDNA that codes for this protein. Bovine adrenal gland and human pheochromocytoma were used as sources of the mRNA for these cloning studies.[36-38] The structures of adrenal proenkephalin A from these two species reveal a high degree of homology and the same precursor is used in brain tissue to generate the enkephalins. Proenkephalin A contains six copies of [Met]-enkephalin and a single copy of [Leu]-enkephalin. These proportions are consistent with the higher brain concentration of [Met]-enkephalin compared with [Leu]-enkephalin.[31] Two of the six copies of the [Met]-enkephalin sequence in proenkephalin A possess C-terminal extensions of two or three amino acids, giving rise to the so-called "octapeptide" [Met]-enkephalin-Arg[6]-Gly[7]-Leu[8]) and the "heptapeptide" [Met]-enkephalin-Arg[6]-Phe[7]). Each of these sequences is sandwiched between pairs of basic amino acid residues important in the processing of the proenkephalin precursor (Fig. 2). The occurrence of both [Met]- and [Leu]-enkephalin within a single precursor suggests that both peptides may be stored in the same neurons. However, it has been demonstrated that some neurons show immunoreactivity to [Leu]-enkephalin but not to [Met]-enkephalin. In these neurons, the [Leu]-enkephalin sequence must arise from expression of the proenkephalin B gene. Proenkephalin B (Fig. 2) contains no [Met]-enkephalin sequence but produces three main [Leu]-enkephalin-containing peptides: α/β-neoendorphin,

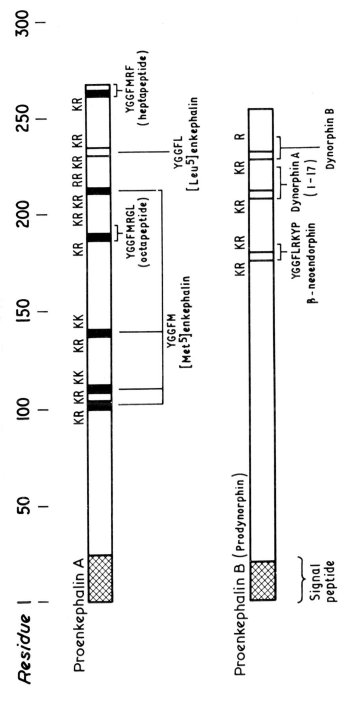

Fig. 2. Schematic diagram of the human preproenkephalin A and preproenkephalin B (preprodynorphin) precursor polypeptides as deduced from cloning and sequencing the corresponding cDNA (see ref. 35). The location of the signal peptide region, the biologically active neuropeptide domains and the dibasic amino acid cleavage recognition sites are indicated. The one-letter code for amino acids is used: K, lysine; R, arginine; C, glycine; L, leucine; M, methionine; F, phenylalanine; P, proline; and Y, tyrosine. See text for further details. ■, YGGFM sequence; □, YGGFL sequence.

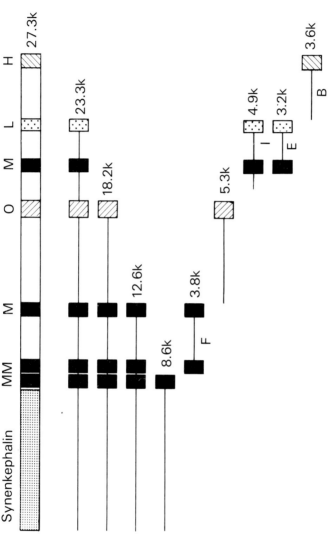

Fig. 3. Proenkephalin and the various enkephalin-containing peptides in the bovine adrenal medulla. The numbers (27·3 k, 23·3 k, 18·2 k, 12·6 k, etc.) refer to the M_r values of peptide fragments. Note that, in contrast to brain, larger proenkephalin-containing peptides predominate in adrenal medulla. Peptide E contains a [Met]-enkephalin sequence at its N-terminus and a [Leu]-enkephalin sequence at its C-terminus. Peptide F has a [Met]-enkephalin sequence at each end. It is, as yet, unclear which peptides are processing intermediates and which are end-products in the adrenal medulla. Synenkephalin is a cysteine-rich non-opioid peptide comprising 70 residues at the N-terminus of proenkephalin. M, [Met]-enkephalin; L, [Leu]-enkephalin; O, octapeptide; H, heptapeptide. (Reproduced from Patey, G., Liston, D. & Rossier, J. (1984). *FEBS Lett.* **172**, 303–308, with permission of the Federation of European Biochemical Societies and the publisher.)

dynorphin A and dynorphin B or rimorphin.[35,39,40] Dynorphin may serve as a precursor of [Leu]-enkephalin in the brain but, in view of its own potent opiate actions, it is likely to have an independent physiological role. A number of other peptides, some with and others without opiate activity, can also be generated from the proenkephalin A and B precursor polypeptides, although the precise cell complement of opioid peptides is tissue-specific.[41] Figure 3 illustrates the range of enkephalin-derived products detectable in adrenal medulla that can arise during processing of proenkephalin A. In general, enkephalin-containing neurons colocalize with opiate receptors and involve brain structures whose functions are known to be linked to opiate actions. The enkephalins and their receptors are also present in the limbic system and may therefore regulate emotional behaviour. As already mentioned, in the adrenal medulla the peptides may act in concert with catecholamines in stress situations. The presence of so many opioid peptides coded for by at least three distinct genes argues for a much broader role for opioid peptides than simply as the body's own analgesic agents.

B. SUBSTANCE P AND THE TACHYKININS

Of all the neuropeptides, substance P has the longest history,[42] being discovered in 1931, and is one of the best characterized so far as distribution and function are concerned (see e.g. ref. 43). It was originally identified[42] in the gut as a hypotensive factor distinct from acetylcholine and named substance P simply on the basis of being isolated as a white powder (P for Pulver). The name has stuck to this day. Although suggested to be a protein by von Euler, its structure was not elucidated until 1970 when it was shown to be a peptide of 11 amino acids.[44] The sequence of substance P is given in Table 3, where it is seen to have features in common with the molluscan salivary gland peptide eledoisin and the amphibian skin peptides such as kassinin and physalamein, generically referred to as tachykinins. These peptides share a common C-terminal sequence, Phe—Xaa—Gly—Leu—MetNH$_2$, and exhibit similar biological activities including reduction of blood pressure, smooth muscle contraction, salivation and stimulation of glandular secretion. Until recently, substance P was the only known mammalian tachykinin where it was believed to function as one of the sensory nerve transmitters involved in the transfer of nociceptive information.[45] However, its role as a "pain transmitter" remains speculative. More recently, two other tachykinins have been described in the mammalian CNS. Confusingly, these novel neuropeptides have been given various names but are now unambiguously termed *neurokinin A* (substance K; neuromedin L;

TABLE 3

The family of tachykinin peptides

Mammalian
Substance P   Arg—Pro —Lys —Pro—Gln —Gln—Phe—Phe—Gly—Leu—MetNH$_2$
Neurokinin A   His —Lys —Thr—Asp—Ser —Phe—Val —Gly—Leu—MetNH$_2$
Neurokinin B   Asp—Met—His—Asp—Phe—Phe—Val —Gly—Leu—MetNH$_2$

Amphibian
Hylambatin   Asp—Pro —Asp—Pro—Asp—Arg—Phe—Tyr—Gly—Leu—MetNH$_2$
Kassinin   Asp—Val —Pro —Lys —Ser —Asp—Gln—Phe—Val —Gly—Leu—MetNH$_2$
Phyllomedusin   Glp—Asn—Pro—Asn—Arg—Phe—Ile —Gly—Leu—MetNH$_2$
Physalaemin   Glp—Ala—Asp—Pro—Asn—Lys —Phe—Tyr—Gly—Leu—MetNH$_2$
Uperolein   Glp—Pro —Asp—Pro—Asn—Ala—Phe—Tyr—Gly—Leu—MetNH$_2$

Molluscan
Eledoisin   Glp—Pro —Ser —Lys —Asp—Ala—Phe—Ile —Gly—Leu—MetNH$_2$

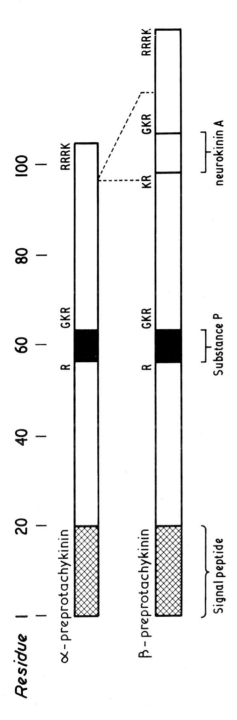

Fig. 4. Schematic diagram of bovine brain α- and β-preprotachykinins. The location of the signal peptide region, the biologically active neuropeptide domains and the cleavage recognition sites are indicated. The one-letter code for amino acids is used: K, lysine; R, arginine; G, glycine. The dotted line indicates the sequence, which includes the neurokinin A region, that is deleted in α-preprotachykinin (adapted from ref. 49).

neurokinin α) and *neurokinin B* (neuromedin K; neurokinin β) and it is likely that they too function as neurotransmitters or neuromodulators. One of these peptides, neurokinin A, was first isolated as a gut-contracting peptide from spinal cord[46,47] and was shown to have an amino acid sequence and pharmacological activity characteristic of the tachykinin family of peptides. At about the same time, Maggio & Hunter[48] reported the existence in spinal cord of abundant kassinin-like immunoreactivity which turned out to be due to the same peptide and which they termed substance K. This peptide is considerably more potent than substance P in some pharmacological tests and it may therefore act through a different class of receptors and have a somewhat different physiological role from substance P. The relationship between substance P and neurokinin A has been revealed through application of recombinant DNA technology.[49,50] Nakanishi and colleagues were able to clone DNA sequences complementary to bovine striatal mRNAs coding for the substance P precursor. Nucleotide sequence analysis of cloned cDNAs has revealed that the bovine brain substance P precursors (Fig. 4) are encoded by at least two different mRNAs. One encodes only the substance P sequence. The other mRNA encodes both the substance P and neurokinin A sequences. These neuropeptide precursors have been referred to as α- and β-preprotachykinins, respectively.[49] The two preprotachykinin mRNAs exhibit an interesting structural relationship. They are completely identical in their 5' and 3' sequences but differ in insertion (or deletion) of the sequence coding for the neurokinin A region. It now appears that the sequence specifying the neurokinin A region is encoded by a discrete genomic segment and that both α- and β-preprotachykinin mRNAs arise from a single gene by alternative splicing events.[50] Relative amounts of the mRNAs differ between tissues, suggesting that the neurokinin A encoding sequence is regulated in a tissue-specific manner.[50] For example, the ratio of β- to α-preprotachykinin is greatest in thyroid, duodenum and small intestine where substance K may have very specific functions. The α-preprotachykinin mRNA is virtually undetectable in the thyroid. The precursor to neurokinin B has yet to be isolated. The individual tachykinin peptides appear to interact with different populations of tachykinin receptors.[51]

C. CALCITONIN AND CGRP

Alternative RNA processing events, as seen in expression of the tachykinin genes, may be relatively common during the expression of many eukaryotic genes. A particular example is the differential expression of the calcitonin gene between thyroid and brain.[15,52] In thyroid "C" cells, the major product is the peptide hormone calcitonin, whereas in brain, alternative RNA splicing leads to the production of a different product referred to as calcitonin gene-related peptide (α-CGRP). An mRNA product of a related gene has been identified in rat brain and thyroid which encodes the protein precursor of a peptide differing from α-CGRP by only a single amino acid.[53] The RNA encoding this peptide, which has been referred to as β-CGRP, appears to be the only mature transcript of the β-CGRP gene. Hybridization histochemistry has shown α- and β-CGRP mRNAs to be similarly distributed, although their relative levels of expression may differ in different regions. Thus α- and β-CGRP represent members of yet another family of related genes encoding peptides with potential functions as neurotransmitters. This family of peptides appears to have functions in controlling sensory and motor information as well as cardiovascular homeostasis.[54]

D. ATRIAL NATRIURETIC FACTORS (ATRIOPEPTINS)

Atrial natriuretic factor (ANF) comprises a family of peptides derived from a single precursor referred to as pronatriodilatin.[55] These peptides were originally identified as the constituents of electron-dense granules present in the atrial muscle of heart.[56] The regulatory peptides have important roles in the regulation of water and salt balance and act on specific receptors located in kidney, adrenal cortex, smooth muscle cells and elsewhere.[57] The discovery of these peptides has led to the concept that the heart, that elicits the hormones, has a physiological role to play as an endocrine organ.[58] The rapid advances in this area, particularly the contributions of recombinant DNA technology, have recently been reviewed (see e.g. refs 57–59).

As with other regulatory peptides originally isolated from peripheral tissues, it comes as no surprise that ANF also occurs in the CNS.[10,60] It has been shown to be widely distributed in the brain by radioimmunoassay and immunohistochemical techniques. The functions of ANF in the brain are unknown but its distribution, with highest concentrations in the hypothalamus and septum, would be consistent with a role in controlling thirst and perhaps the release of vasopressin. The molecular forms of ANF in brain and heart differ with low molecular weight forms predominating in the brain. This suggests that there is tissue-specific processing of the precursor peptides, which may reflect different receptor populations and functional roles for the peptide products in different tissues. As will be apparent from all the examples that have been described, this phenomenon appears to be a common theme among neuropeptide families.

III. Processing of Neuropeptide Precursors

A. ORGANIZATION OF NEUROPEPTIDE BIOSYNTHESIS

Most of the brain peptides examined to date (the dipeptide carnosine (β-alanyl histidine) being an exception) appear to be synthesized ribosomally as large precursor proteins which are subjected to post-translational modifications, particularly limited proteolysis. Thus, those neurons utilizing peptides as neurotransmitters ("peptidergic neurons") must differ in their structural organization from neurons utilizing classical transmitters alone.[14,23,61] These differences are illustrated schematically in Fig. 5. For example, the catecholamines (dopamine, noradrenaline), acetylcholine and GABA are all synthesized in the nerve terminal and the released transmitter can be replaced either by new synthesis or by re-uptake of intact transmitter from the synapse. In these cases transmitter synthesis occurs predominantly, if not exclusively, in the terminal region of the neuron.

In the case of peptides, their synthesis occurs only on ribosomes in the cell body (perikaryon) region and there is no local synthesis at the nerve terminal, which lacks ribosomes. Processing to form the mature, biologically active peptide probably occurs in the storage vesicles during axonal transport from cell body to nerve terminal. As no re-uptake mechanisms occur for neuropeptides, each peptide molecule that is released can only be replaced through new protein synthesis and axonal transport. This relatively slow and inefficient mechanism for replenishing the transmitter suggests that peptides may act intermittently with a relatively long duration of action.

The organization of neuropeptide biosynthesis, as for all secreted proteins, begins on the rough endoplasmic reticulum where the initial translation product (the preprotein) is formed and, guided by the "signal peptide" at the N-terminus, is channelled into the cisternae of the endoplasmic reticulum (see e.g. ref. 62). Cleavage of the hydrophobic "pre" or signal sequence occurs rapidly, probably as a cotranslational event while the nascent peptide chain enters the cisternae. The enzyme that catalyses removal of the signal peptide ("signal peptidase") appears to be a metalloendopeptidase located on the inner membrane of the rough e.r.[63] The signal peptidase has defied purification for some years. However, Blobel's group has recently isolated the enzyme from dog microsomes as a complex composed of six polypeptides, two of which are glycosylated.[63]

After the removal of the signal peptide, further proteolytic processing is generally required before the active regulatory peptide is completed.[64,65] A combination of biochemical, electron microscopic and autoradiographic evidence suggests that the proprotein is translocated from the rough e.r. cisternae to the Golgi apparatus where it is packaged into the secretory

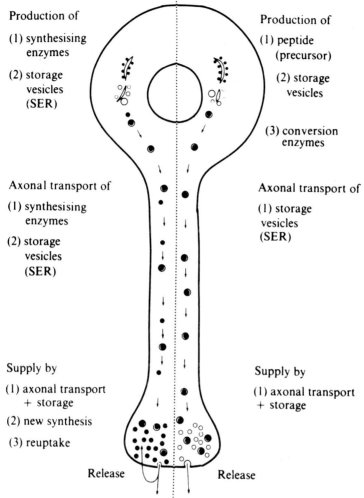

'CLASSICAL' TRANSMITTER   PEPTIDE TRANSMITTER

Production of

(1) synthesising
    enzymes

(2) storage
    vesicles
    (SER)

Production of

(1) peptide
    (precursor)

(2) storage
    vesicles

(3) conversion
    enzymes

Axonal transport of

(1) synthesising
    enzymes

(2) storage
    vesicles
    (SER)

Axonal transport of

(1) storage
    vesicles
    (SER)

Supply by

(1) axonal transport
    + storage

(2) new synthesis

(3) reuptake

Supply by

(1) axonal transport
    + storage

Release   Release

Fig. 5. Schematic diagram of a neuron demonstrating some differences between a neuron using a "classical" transmitter such as noradrenaline (left) and a peptide transmitter (right). SER, smooth endoplasmic reticulum. (Reproduced from ref. 23 with permission of Macmillan and the authors.)

granules ("synaptic vesicles"). For some proproteins, the initial proteolytic cleavage may occur in the Golgi and the final steps then take place within the granules. Other proproteins appear to be processed entirely after packaging into the granules. The first step in the processing normally requires the action of an endopeptidase, although there are some examples, e.g. synthesis of the bee venom peptide, melittin, where an aminopeptidase (dipeptidyl aminopeptidase IV) is the activating enzyme.[66]

The classical work of Docherty & Steiner[67] on the processing of proinsulin demonstrated that a sequence of two basic amino acids serves as the signal to initiate endopeptidase attack. Removal of these basic residues generates the free peptide and involves the consecutive action of two separate classes of enzyme. Since a combination of trypsin and carboxypeptidase B *in vitro* was able to generate insulin and the C-peptide from proinsulin, the enzymes responsible for these processing steps have generally been regarded as being similar to trypsin and carboxypeptidase B. For more complex precursors, multiple cleavage sites must occur. For example, to release all of the [Met]- and [Leu]-enkephalin sequences from the proenkephalin A precursor requires hydrolysis at 13 separate cleavage sites (Fig. 2). It is now becoming clear that there may exist a unique set of processing enzymes localized in secretory granules that carry out these processes.[64,65] Only very recently have attempts begun to isolate and characterize these enzymes. The problem to date has been to distinguish such enzymes which appear to be present in low abundance compared with similar proteases located elsewhere in the cell, particularly in lysosomes.

B. ENDOPEPTIDASE CLEAVAGE

The initial event in processing of prohormone is the endopeptidase attack at the basic (either lysine or arginine) residues flanking the neuropeptide sequence. Usually the carboxy-terminal amino acid in this pair is arginine. Various approaches have been used to identify the proteases responsible for these processing events. Many different proteolytic enzymes are potentially capable of effecting the cleavages. As yet, there appears to be no consensus as to the identity of the enzyme(s) responsible for the trypsin-like cleavage, but one possibility has been that a member of the kallikrein family may be involved.[68] These are serine proteases with trypsin-like specificity that are already known to be responsible for the processing of epidermal growth factor and nerve growth factor from their precursors.[69,70] In these unusual cases, the processing enzymes are cosynthesized with the growth factor precursor on an equimolar basis. Further evidence for the involvement of one or more serine proteases in pro-polypeptide conversion has come from outside the neuropeptide field. In a recently reported case of human proalbuminaemia, the deficiency, was due to a mutant α_1-antitrypsin inhibitor.[71] The mutation converted the inhibitor from an anti-elastase to an anti-thrombin specificity, suggesting that the proalbumin-converting enzyme might be closely related to thrombin which, like the kallikreins, is an arginyl-endopeptidase.[72] On this basis Green has argued[73] that prohormone activating enzymes could form part of a regulatory enzyme cascade.

In addition to the kallikreins, a thiol protease has been implicated in the processing of proinsulin to insulin in pancreatic islets,[74] POMC to ACTH in the pituitary[75] and the formation of the enkephalins in the adrenal medulla.[76] The pancreatic enzyme was similar to the lysosomal protease cathepsin B and may therefore be of lysosomal origin. The pituitary and adrenal proteases, however, appear to be associated with secretory granules and to be distinct from cathepsin B. The confusion over enzyme identity that proliferates in this area and the multiplicity of potential processing candidates is probably a reflection of the fact that several such classes of processing enzyme exist, perhaps with different tissue distributions and specificities. The development of model substrates and selective enzyme inhibitors may be helpful in identifying the physiologically relevant processing enzymes. Alternative approaches worth exploiting are the use of cell lines or more primitive organisms producing regulatory peptides. Kelly and colleagues[77] have been able to show that AtT20 cells, a mouse anterior pituitary cell line, which are normally able to express the POMC gene and process the POMC precursor, can also correctly process proinsulin if the proinsulin gene is transferred into these cells and subsequently expressed. These particular cells therefore have the capacity to process more than one prohormone and may provide a useful model system for further study. A different approach has been employed by Thorner and collaborators[78] who have used a mutant strain of yeast which was unable to produce the two yeast regulatory peptides, α-factor and killer-toxin. The mutation appeared to be in a gene coding for a processing enzyme which was subsequently characterized as showing specificity for pairs of basic residues. The enzyme was shown to be a specific endopeptidase that cleaved on the carboxyl side of the pair of basic residues. This enzyme may well be similar to the corresponding mammalian processing enzymes, as yet uncharacterized. However, in the case of dynorphin-B, cleavage appears to occur on the amino-terminal side of the arginine of the Lys–Arg pair, the arginine then selectively being removed by an aminopeptidase with specificity for basic residues. That conformation of the precursor peptide may be an important factor determining processing is reflected in the fact that processing by no means occurs at all available dibasic sequences. For example, some biologically active peptides (e.g. ACTH) retain an internal pair of basic residues. Ultimately, characterization of the individual processing enzymes should be able to explain these differences. Site-directed mutagenesis of neuropeptide genes in the vicinity of potential processing sites should provide valuable information on the precise specificity of the processing enzymes.

Although release of a neuropeptide from its precursor often involves hydrolysis at a Lys–Arg pair of residues, there are examples where cleavage occurs at a single basic residue, usually an arginine. At first sight there appears to be no "consensus sequence" of amino acids flanking the single basic residue that might provide a signal for cleavage. However, Schwartz[79] has recently observed that Pro residues frequently occur immediately before or after the susceptible basic residue. He has suggested[79] that the conformation of the peptide backbone induced by the proline may be important in making the single arginine accessible to the processing enzyme. There are far fewer cases where processing has been detected after a single lysine residue. One example[80] is the cholecystokinin (CCK) precursor which can be processed to generate CCK-22. Schwartz[79] has argued convincingly that the monobasic and dibasic processing mechanisms are the result of different processing enzymes expressed on a tissue-specific basis.

C. CARBOXYPEPTIDASE B-LIKE ACTIVITY ("ENKEPHALIN CONVERTASE")

The cleavage of a neuropeptide precursor adjacent to a basic residue releases a peptide with Lys or Arg at the C-terminus. The removal of this basic residue requires the action of an enzyme that has been described as carboxypeptidase B-like. Carboxypeptidase B (protaminase; EC 3.4.17.2) is a zinc metallopeptidase of M_r 34 000 originally isolated from pancreas, that preferentially releases C-terminal Lys or Arg residues. The processing enzyme, which is presumably located in the secretory granules, appears to be distinct from a lysosomal carboxypeptidase B.

Attempts to isolate the carboxypeptidase involved in neuropeptide processing have principally focused on the formation of enkephalin from its arginine-extended precursor, enkephalin-Arg[6], in chromaffin granules from the adrenal medulla.[81,82] A simple assay for the carboxypeptidase has been devised using dansyl–Phe–Leu–Arg as substrate and quantifying the reaction product (dansyl–Phe–Leu) fluorimetrically.[83] Chromaffin granules were shown to contain a unique carboxypeptidase that differs in several characteristics from other carboxypeptidase B-like enzymes and has been designated "carboxypeptidase E" or "enkephalin convertase", reflecting its presumed role as the final processing enzyme in generating biologically active enkephalin.[84] This enzyme has proved more amenable to purification than the endopeptidase processing enzyme and has now been isolated in homogeneous form.[85] Its characteristics are summarized in Table 4. Purification was largely achieved through affinity chromatography on Arg-Sepharose resulting in over 100 000-fold enrichment from brain. Both membrane-bound and soluble forms of the enzyme exist, as is the case for the noradrenaline biosynthetic enzyme, dopamine β-hydroxylase. The

TABLE 4

Properties of the secretory granule carboxypeptidase processing enzyme (enkephalin convertase or carboxypeptidase E)

Specificity	—○—○—○—● where ● is a basic residue
Major locations	Adrenal medulla, pituitary, brain
M_r	50 000
pH optimum	5–6
Activators	Co^{2+}
Inhibitors	Chelating agents (1,10-phenanthroline, EDTA); GEMSA
Active site	Zn^{2+}

membrane-bound form is somewhat larger than the soluble form, probably due to the presence of a hydrophobic "tail" that anchors the enzyme in the chromaffin granule membrane.

In the pituitary, a single carboxypeptidase is able to process correctly a variety of different pituitary hormone precursors and therefore a term such as "enkephalin convertase", implying a unique specificity for enkephalin precursors, is clearly a misnomer. It has, however, been all too common in the literature for enzymes to be named solely on the basis of the first substrate with which they are assayed, before their wider specificity is recognized. "Angiotensin-converting enzyme" and "enkephalinase" provide two other such cautionary examples from the neuropeptide field (see below).

The localization of the carboxypeptidase processing enzyme has been facilitated by the discovery that guanidinoethylmercaptosuccinate (GEMSA) is a potent (K_i = 8 nM) and selective inhibitor of this particular enzyme.[86] Since [³H]-GEMSA apparently binds exclusively and with high affinity to a single protein in brain, the labelled inhibitor can be used to localize the carboxypeptidase autoradiographically.[87] Of all the tissues examined, the highest levels of the enzyme are found in the anterior pituitary, which contains very little enkephalin, further arguing against an exclusive role as an enkephalin biosynthetic enzyme. Fricker[84] has drawn attention to the fact that the correlation between overall neuropeptide biosynthesis and carboxypeptidase-E levels is much closer than the correlation with any individual neuropeptide. Thus it is likely that a single carboxypeptidase-processing enzyme is involved in the production of many peptide hormones and neurotransmitters. Indeed, high specificity of such processing is not essential, as Fricker[84] has pointed out, since specificity is ensured by packaging together the peptide precursors and the carboxypeptidase within secretory granules. Thus it may not prove possible to interfere with neuropeptide biosynthesis selectively by pharmacological inhibition of the carboxypeptidase.

D. C-TERMINAL AMIDATION

Many regulatory peptides are protected from hydrolysis at the C-terminus by amidation and the presence of this amide group generally appears to be essential for biological activity. Examples include substance P and other tachykinins, LH-RH, thyroliberin (TRH) and α-MSH. Bradbury et al.[88] were able to show that a C-terminal glycyl residue in the precursor serves as amino-donor in the amidation reaction. In all cases for which the precursor sequence has been established, the active peptide sequence is immediately followed by Gly and then one or two basic residues. The amidating enzyme, purified from pituitary, can be assayed using a "model" substrate, D-Tyr—Val—Gly, which is converted to D-Tyr—ValNH$_2$. The mechanism of the reaction appears to be:

$$\text{D-Tyr—Val—Gly} + O_2 \rightarrow \text{D-Tyr—ValNH}_2 + \text{glyoxylate} + H_2O$$

Ascorbate is required as a cofactor for this reaction and impaired amidation of the peptide hormone gastrin has been shown to occur in ascorbate-deficient guinea-pigs.[89] In the AtT-20 cell line, the amidating enzyme, which is located in secretory granules, is secreted together with the POMC-derived peptides in response to corticotropin-releasing factor (CRF) or cyclic AMP. Glucocorticoids and dopamine analogues can bring about changes in amidating activity, suggesting that regulation of this processing enzyme might be a key control point in the synthesis of many biologically active peptides.[90]

E. OTHER MODIFICATION PROCESSES

Many biologically active peptides are protected at the N-terminus from the action of aminopeptidases and a number of peptides, e.g. LH-RH and TRH are protected at both termini. Where glutamate is the N-terminal residue, protection involves cyclization to form a pyroglutamyl group, as seen in LH-RH, neurotensin and TRH. The modification of an N-terminal residue of a peptide can also occur by acetylation, as is seen with α-MSH and β-endorphin. A specific endorphin acetyltransferase has been characterized from rat brain and pituitary and may also function in the acetylation of α-MSH.[91] Whereas acetylation of α-MSH markedly increases its biological activity, the acetylated form of β-endorphin has negligible analgesic activity and little affinity for the opiate receptor.[92] A number of other post-translational modification reactions are known, including methylation, phosphorylation and tyrosyl O-sulphation. Cholecystokinin, for example, can exist in sulphated forms. The sulphation enzyme appears to be a membrane-bound Golgi protein[93] distinct from the cytosolic phenolsulphotransferase involved in sulphation of catecholamine metabolites.

F. THE PROCESSING OF NEUROPEPTIDE PRECURSORS IS TISSUE-SPECIFIC

Although the enzymes involved in peptide processing have yet to be all fully characterized, it is clear that the generation of a biologically active peptide from its precursor proceeds through a limited and controlled proteolysis catalysed by the specific processing peptidases (converting enzymes), which are predominantly localized in the secretory vesicles. However, the same primary translation product (polyprotein) can be processed to distinct end-products in different tissues, thus allowing diversity of function of neuropeptide gene products. The best studied example (Fig. 6) is provided by POMC, which is post-translationally processed to ACTH in the anterior lobe of the pituitary and to MSH in the intermediate lobe.[94] The precursors for cholecystokinin and proenkephalin can also be processed differently in different cell types, reinforcing the concept of multiple processing enzymes that can be expressed in a tissue-specific manner.[95,96] This is clearly the case for the enzyme involved in N-terminal acetylation of MSH which is found exclusively in MSH cells. The pattern of processing of a multi-functional prohormone may also vary with physiological state, e.g. in response to hormones, neurotransmitters or diurnal variations. In the anterior lobe of the pituitary, the secretion of POMC-derived peptides is regulated positively[97] by hypothalamic corticotropin releasing factor (CRF) and negatively by glucocorticoids. In the intermediate lobe, dopamine can inhibit the release of the peptides and the dopamine antagonist haloperidol causes a five-fold increase in the levels of POMC mRNA.[98] Haloperidol can also selectively increase the [Met]-enkephalin content in striatum and it may be that the beneficial effects of haloperidol treatment in schizophrenia relate, in part, to the increased production of striatal enkephalins.[99]

The intricate possibilities for the processing of multi-functional prohormones have been likened by Smyth[100] to the orchestration of musical sounds:

> Amplification of individual notes is accompanied by the dampening of others and exquisite variation made possible by the order in which the notes are played. In multifunctional pro-hormones, it appears that a variety of different processing reactions are employed to produce complex physiological effects. The subtle control of these reactions is likely to reside in the regulation of the specific processing enzymes.

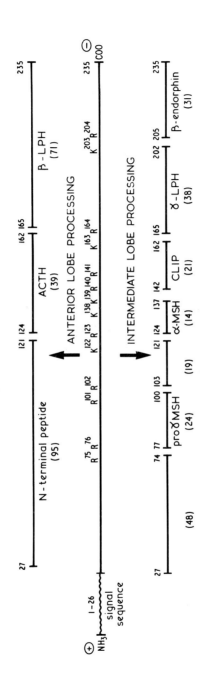

Fig. 6. Tissue-specific processing of mouse pro-opiomelanocortin in the anterior and neurointermediate lobes of the pituitary. The various biologically active domains of the peptide are noted as well as the basic amino acid residues (K, lysine; R, arginine) that signal processing sites. (Reproduced from ref. 94 with permission of the authors.)

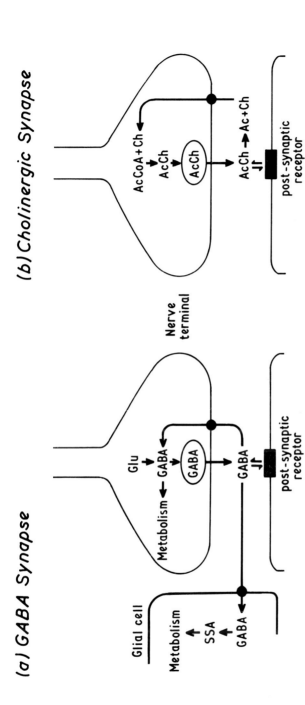

Fig. 7. Mechanisms of neurotransmitter removal at (a) GABA and (b) cholinergic synapses. GABA is removed by Na⁺-dependent uptake into neuronal and glial cells followed by metabolism. Acetylcholine (AcCh) is hydrolysed extracellularly to acetate (Ac) and choline (Ch) by membrane-bound acetylcholinesterase. The choline can be taken up into neuronal cells and re-used to form acetylcholine.

IV. Inactivation of Neuropeptides

A. MODES OF NEUROTRANSMITTER INACTIVATION

The concept of the post-synaptic inactivation of neuropeptides has developed largely by analogy with events occurring at synapses employing classical neurotransmitters, typically acetylcholine. Two contrasting mechanisms of neurotransmitter removal are indicated in Fig. 7. At the GABA synapse, the action of GABA is terminated by its uptake intact into both neuronal and glial cells.[101] The transport protein for this process is located in the plasma membrane and is Na^+- and Cl^--dependent. The protein has recently been partly purified[102] and, when reconstituted into liposomes, can mediate GABA transport in the presence of the appropriate ions. The re-absorbed GABA is then metabolized intracellularly, initially to succinic semialdehyde (SSA), by mitochondrial GABA-transaminase. In the case of monoamine synapses there is competition between uptake by nerve terminals and glial cells. After pre-synaptic uptake the amine may either be transported into storage vesicles for re-use or inactivated by monoamine oxidase. After glial cell uptake inactivation involves the action of monoamine oxidase and, in the case of catecholamines, catechol-O-methyltransferase.

At the cholinergic synapse, released acetylcholine is hydrolysed extracellularly by acetylcholinesterase (Fig. 7), the choline so formed subsequently being re-used to form more acetylcholine. Acetylcholinesterase is an ectoenzyme, that is an integral membrane protein of the plasma membrane with its active site exposed on the extracellular face of the membrane. A similar topological organization appears to occur for the enzymes that hydrolyse released neuropeptides.

B. ENKEPHALIN METABOLISM

Investigations of neuropeptide metabolism inevitably began with the enkephalins as peptide substrates and it was soon established that [Leu[5]]- and [Met[5]]-enkephalin were rapidly hydrolysed by tissue preparations to inactive metabolites.[103,104] Spurred on by the possibility of developing selective inhibitors of peptide metabolism, this research has progressed to a detailed molecular characterization of enkephalin-degrading peptidases. Much of the original philosophy behind this work has been based on the possibility that there might exist neuropeptide-specific peptidases ("neuropeptidases", see e.g. refs 105, 106) and has led to the use of nomenclature such as "enkephalinase", "Substance P degrading enzyme" and similar terms implying a selectivity of peptidase action.[105–107] Clearly such specificity would be welcome to the pharmacologist and was a distinct possibility when only a few neuropeptides were recognized. This concept now appears untenable since it would require an elaborate collection of peptidases to deal with the tens of neuropeptides currently recognized. An alternative, and more economical arrangement, is that there exist a limited array of peptidases in the plasma membrane of many different cell types and it is their localization rather than peptide specificity that is the critical factor in terminating peptide action.

C. MICROVILLAR PEPTIDASES AS A MODEL SYSTEM

In our investigations of the metabolism of neuropeptides, we have consistently held the view that synaptic membrane peptidases are unlikely to be enzymes unique to the nervous system.[108-110] Other tissues or species enriched in membrane peptidases could therefore provide a useful starting point for isolation and characterization of the peptidases of the synapse. Analogies may be drawn with other neurotransmitters. Monoamine oxidases, for example, not only degrade catecholamines and indoleamines, but also oxidize a wide range of aromatic and aliphatic amine substrates in brain, liver and other tissues. The initial isolation of the nicotinic acetyl-choline receptor, and the molecular cloning of the genes for the receptor subunits, took advantage of the enrichment of this protein in the electrop-laques of electric fishes such as Torpedo. Only subsequently was the mammalian receptor isolated and characterized.

The plasma membrane peptidases best characterized in molecular terms are undoubtedly those located in the brush borders of kidney and intestine (see e.g. refs 111, 112 for reviews). Some of those that have been charac-terized in the kidney microvillar membrane are listed in Table 5. These enzymes, although highly enriched in the kidney brush border, are not unique to the proximal tubule cell. A comparable assembly of peptidases is also present in abundance in intestinal microvilli. The combined specificities of the enzymes provide a battery of efficient peptidases that can degrade small peptides to their constituent amino acids. Selective and potent inhibitors are available for some of these peptidases, several of which have been purified and characterized in considerable detail.[110,112] Knowledge of the microvillar enzymes can therefore provide a useful model system for investigation of membrane peptidases from other tissues. More importantly, our collection of monoclonal and polyclonal antibodies to renal peptidases has proved a valuable experimental tool for the isolation and characteriz-ation of synaptic peptidases.[113,114]

The microvillar peptidases listed in Table 5 are all integral membrane glycoproteins sharing a number of structural similarities.[112] The bulk of each

TABLE 5

Peptidases of the pig kidney microvillar membrane

Enzyme	$M_r \times 10^{-3}$	Active site	Specificity	Inhibitors	Peripheral	Nervous system
					Main locations	
Endopeptidase-24.11 ("enkephalinase")	90	Zn^{2+}	(● = hydrophobic)	Phosphoramidon, thiorphan	Kidney, lymph nodes, intestine, glands	Choroid plexus Striatum Substantia nigra and peripheral nervous system
Peptidyl dipeptidase A (angiotensin converting enzyme)	180	Zn^{2+}	(many)	Captopril, enalaprilat, lisinopril	Kidney, lung	Choroid plexus Striatum Subfornical organ
Aminopeptidase N	160	Zn^{2+}	(many)	Amastatin, bestatin, actinonin	Kidney, intestine lung	Widely distributed
Aminopeptidase W	130	Zn^{2+}	(● = Trp)	Amastatin	Kidney, ileum	Detectable in Striatum
Dipeptidyl peptidase IV	130	Serine	(● = Pro)	Diisopropylfluoro-phosphate	Kidney, intestine	?
Aminopeptidase A	170	Ca^{2+}	(● = Glu/Asp)	Amastatin	Kidney,	?

The data are for the enzymes from pig kidney. Enzymes located from other tissues may differ slightly in M_r due to differences in the extent of glycosylation enzyme is hydrophilic, with only a small hydrophobic region (less than 5% of the mass of the protein) that serves to anchor it in the membrane. The active sites are located on the hydrophilic portion of the enzymes facing the extracellular space. If present in the nervous system, therefore, the enzymes would be oriented appropriately for the hydrolysis of released peptide.

D. SYNAPTIC "ENKEPHALINASE" IS IDENTICAL WITH MICROVILLAR ENDOPEPTIDASE-24.11

A major metabolite of [Met]-enkephalin in the perfused brain *in vivo*[115] is Tyr—Gly—Gly which arises by hydrolysis of the Gly^3—Phe^4 bond of the enkephalin, releasing the dipeptide Phe–Met. Both fragments are physiologically inactive. Subsequent studies *in vitro* have confirmed that preparations from the striatum, a brain region rich in peptidergic nerve terminals, contain a membrane enzyme that is able to hydrolyse the Gly^3—Phe^4 bond of enkephalin.[108,116] This enzyme (colloquially termed "enkephalinase") has been the focus of much attention by neurochemists, since inhibition of the enzyme could potentiate enkephalin action. "Enkephalinase" inhibitors might therefore provide a novel class of analgesic agents acting at a distinct target from the opiate analgesics such as morphine. Current progress and setbacks in this direction have recently been reviewed.[117,118] The identity of "enkephalinase" was a matter of considerable speculation and confusion in the five years after its initial discovery in the striatum in 1978, in part because it was assumed to be brain- and peptide-specific. When a similar activity was detected in other tissues and was shown to be particularly enriched in renal tissue, its identity with the only plasma membrane endopeptidase in pig kidney (now named endopeptidase-24.11, reflecting its enzyme classification as EC 3.4.24.11) was quickly established using specificity, inhibitor and immunological criteria.[108,109,113,119,120] Confirmation of identity[113] came after purification of the brain enzyme to homogeneity using a monoclonal antibody raised to the kidney enzyme (Fig. 8). The specificity of renal and brain enzymes is for hydrolysis on the amino-terminal side of a hydrophobic residue (e.g. Phe, Tyr, Leu, Val) and the enzyme therefore has the potential to hydrolyse a wide range of regulatory peptides which generally contain at least one internal hydrophobic residue. Table 6 illustrates the sites of cleavage of a number of peptide substrates and emphasizes the broad potential of this ectoenzyme to terminate the signals generated by a wide range of regulatory peptides, be they neuropeptides, peptide hormones or immunoregulatory peptides.[110] Although the enkephalins may be important physiological substrates for the enzyme, evidence is accumulating that the tachykinin peptides such as substance P may also be inactivated by this enzyme in the brain.[109,121,122] Of all the peptides examined with endopep-

TABLE 6

Primary sites of hydrolysis of peptides by endopeptidase-24.11

Opioid peptides

[Met]-enkephalin   Tyr—Gly—Gly↓Phe—Met

[Leu]-enkephalin   Tyr—Gly—Gly↓Phe—Leu

[Met]-enkephalin-Arg[6]—Phe[7]   Tyr—Gly—Gly↓Phe↓Met—Arg—Phe↓

Dynorphin-(1-9)-peptide   Tyr—Gly—Gly↓Phe—Leu—Arg—Arg—Ile—Arg

α-Neo-endorphin   Tyr—Gly—Gly—Phe—Leu—Arg—Lys—Tyr—Pro—Lys

Tachykinins

Substance P   Arg—Pro—Lys—Pro—Gln—Gln↓Phe↓Phe—Gly↓Leu—MetNH₂

Neurokinin A   His—Lys—Thr—Asp—Ser↓Phe—Val—Gly↓Leu—MetNH₂

Neurokinin B   Asp—Met—His—Asp↓Phe—Phe—Val—Gly↓Leu—MetNH₂

Physalaemin   Glp—Ala—Asp—Pro—Asn—Lys—Phe—Tyr—Gly↓Leu—MetNH₂

Other peptides

Cholecystokinin-8(sulphated) Asp—Tyr(SO$_3$H)—Met—Gly—Trp—Met—Asp—PheNH$_2$

FMRFamide Phe—Met—Arg—PheNH$_2$

Bradykinin Arg—Pro—Pro—Gly—Phe—Ser—Pro—Phe—Arg

Gastrin releasing peptide Gly—Asp—His—Trp—Ala—Val—Gly—His—Leu—MetNH$_2$

Neurotensin Glp—Leu—Tyr—Glu—Asn—Lys—Pro—Arg—Arg—Pro—Tyr—Phe—Leu

Chemotactic peptide fMet—Leu—Phe

The symbols ↓ and ⇣ represent the sites of cleavage under initial rate conditions and those occurring in addition after extensive substrate hydrolysis, respectively.

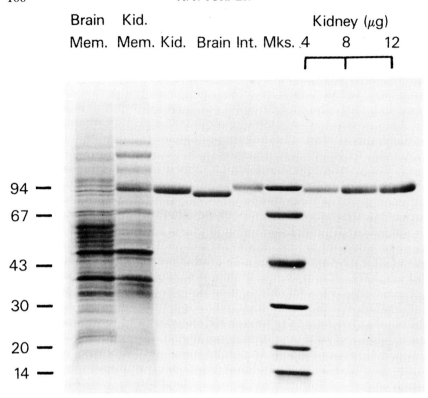

Brain   Kid.   Kidney (μg)
Mem.   Mem.   Kid.   Brain   Int.   Mks.   4   8   12

94 —
67 —
43 —
30 —
20 —
14 —

Fig. 8. SDS/polyacrylamide gel electrophoresis of endopeptidase-24.11 from various sources. The gradient gel comprised 7–17% (w/v) polyacrylamide. Tracks (from left to right): Brain Mem., brain membrane preparation (100 μg); Kid. Mem., kidney microvillar membrane preparation (100 μg); Kid., kidney endopeptidase-24.11 (10 μg); Brain, brain endopeptidase-24.11 (10 μg); Int., intestinal endopeptidase-24.11 (10 μg); Mks, marker proteins (phosphorylase, serum albumin, ovalbumin, carbonic anhydrase, trypsin inhibitor, α-lactalbumin). Kidney (μg), 4, 8 and 12 indicate these amounts of kidney endopeptidase-24.11 run (from ref. 113). Reprinted with permission from The Biochemical Society, London.

tidase-24.11, the tachykinins are hydrolysed the most efficiently, considerably more so than are the enkephalins.[123] Furthermore the hydrolysis of tachykinins by synaptic membranes is abolished in the presence of inhibitors of endopeptidase-24.11. Finally, immunohistochemical studies have shown that the enzyme is particularly abundant in those areas of the brain that are rich in substance P, and particularly so in a major striato-nigral pathway.[124] However, the precise cellular location of the enzyme, whether neuronal or glial, has yet to be confirmed. Ultimately, biochemical investigations alone cannot establish the physiological roles of an individual enzyme, which will require, in addition, a combination of physiological and pharmacological studies. What has been established by the biochemical work is that endopeptidase-24.11 has the specificity and localization to play an important role in terminating peptidergic signals in the nervous system and elsewhere. Its major presence in renal membranes, where it is 5% of the microvillar membrane protein, justifies the use of microvillar peptidases as a model system rich in peptidases for more general studies of peptide hydrolysis at mammalian cell surfaces.

E. AMINOPEPTIDASES CAN INACTIVATE ENKEPHALINS AND OTHER SUSCEPTIBLE NEUROPEPTIDES

Inactivation of the enkephalins can also occur through the hydrolysis of the $Tyr^1–Gly^2$ bond by aminopeptidase action.[103] Much of the aminopeptidase activity in brain is cytosolic and therefore not relevant to the synaptic metabolism of the enkephalins. However, significant quantities of membrane-bound aminopeptidase also occur in brain and this activity is comparable to or exceeds that of endopeptidase-24.11 in all brain regions examined. Since the K_m values for the enkephalins are of the same order of magnitude for aminopeptidase and endopeptidase, it would appear likely that the combined attack of both enzymes is responsible for the extracellular metabolism of the enkephalins. Some support for this contention comes from the observation that inhibition of both endopeptidase-24.11 (with thiorphan) and aminopeptidase (with bestatin) provides greater potentiation of enkephalin-induced analgesia than is produced by inhibition of either enzyme alone.[125] Furthermore a combination of bestatin with thiorphan provides maximal protection from metabolism of endogenous enkephalin released from brain slices by depolarization.[126] The identity of "enkephalin aminopeptidase" has also been a contentious question. Again, however, the renal microvillar membrane system has been the key to establishing that a brain membrane-bound "enkephalin aminopeptidase" is immunologically and structurally identical with kidney aminopeptidase N, an established plasma membrane ectoenzyme.[114,127] A number of neuropeptides in addition to enkephalin have an unblocked N-terminus (e.g. cholecystokinin, neurokinin B) and might therefore be susceptible to hydrolysis and inactivation by membrane aminopeptidases in addition to endopeptidase attack.

F. ANGIOTENSIN CONVERTING ENZYME (PEPTIDYL DIPEPTIDASE A) IS ALSO COMMON TO RENAL AND SYNAPTIC MEMBRANES

Of the characterized cell-surface peptidases, angiotensin converting enzyme (ACE or more properly peptidyl dipeptidase A, EC 3.4.15.1) is

TABLE 7

Primary sites of hydrolysis of some biologically active peptides by angiotensin converting enzyme

Acting as a peptidyl dipeptidase

Angiotensin	Asp—Arg—Val—Tyr—Ile—His—Pro—Phe$\rightarrow$His—Leu
Bradykinin	Arg—Pro—Pro—Gly—Phe—Ser—Pro—Phe$\rightarrow$Arg
Neurotensin	Glp—Leu—Tyr—Glu—Asn—Lys—Pro—Arg—Arg—Pro—Tyr$\rightarrow$Ile—leu

Acting as an endopeptidase

Substance P	Arg—Pro—Lys$\rightarrow$Pro—Gln—Gln—Phe—Phe$\overset{\downarrow}{\rightarrow}$Gly—Leu—MetNH$_2$
LH-RH	Glp—His—Trp—Ser—Tyr—Gly—Leu$\rightarrow$Arg—Pro—GlyNH$_2$

The symbols $\rightarrow$ and $\overset{\downarrow}{}$ represent the sites of cleavage under initial rate conditions and those occurring in addition after extensive substrate hydrolysis, respectively.

perhaps the most intensively studied and inhibitors of the enzyme have been employed successfully as anti-hypertensive agents.[128] There is a very extensive literature on the structure and functions of this enzyme (see e.g. refs 128, 129), and here I shall restrict myself to its potential roles in the brain. The primary specificity of ACE is to act as an exopeptidase releasing successive dipeptide fragments from the C-terminus of an oligopeptide. Its classical physiological substrates are angiotensin I and bradykinin. Recently it has been shown to be present in cerebral tissue where it has been implicated in a brain renin–angiotensin system.[130] In addition, the enzyme may play a role in the hydrolysis and inactivation of C-terminally extended enkephalin peptides,[131] neurotensin[132] and, unexpectedly, substance P.[133–135] In the latter case ACE releases the C-terminal tripeptide fragment from this C-terminally amidated peptide and must therefore function as an endopeptidase with this substrate. LH-RH, which is blocked at both N- and C-termini, can also be hydrolysed anomalously by ACE[136] (Table 7). No adequate model of the action of the enzyme can yet explain these different modes of catalysis by a single peptidase. Its potential for hydrolysis of other biologically active peptides has yet to be explored fully. However, its much wider specificity than originally envisaged, coupled with its relatively high concentration in peptidergic brain regions such as striatum,[137] suggests that it may act in concert with the other brain membrane peptidases in terminating the actions of certain neuropeptides.

G. SUMMARY

It is clear that we do not yet have adequate information to understand fully how neuropeptides are inactivated. There is little, if any, evidence to support a direct uptake mechanism analogous to that for GABA, for removal of neuropeptides from the synapse. However, the possible endocytosis of neuropeptide–receptor complex followed by lysosomal hydrolysis of the adsorbed peptide has yet to be explored. Most research, by analogy with acetylcholinesterase, has concentrated on synaptic hydrolysis of released peptide as the mechanism for peptide inactivation. In the case of the enkephalins, at least two membrane peptidases (endopeptidase-24.11 and aminopeptidases) appear to contribute to its extracellular hydrolysis. However, the relatively high K_m values of the peptidases[123] (typically in the range 10–100 μM) compared with the nanomolar K_D values for neuropeptide receptors suggest that the duration of action of the peptides may be relatively long. Angiotensin converting enzyme may additionally contribute to the hydrolysis of certain neuropeptides in the brain. McKelvy and colleagues[14] have argued that there is now little to support the concept of specific peptide hydrolases tailored for individual neuropeptides. Rather, as we have suggested, evidence favours the existence of a battery of exo- and endopeptidases exhibiting broad specificity located on the surfaces of many different cell types, of which the kidney microvillus is a typical example. It is therefore the location of cells expressing a given peptidase that is the important factor in determining the specificity and relevance of extracellular hydrolysis of neuropeptide signals. Approaches that can be used to identify relevant peptidases include the ability of specific peptidase inhibitors to increase the survival of peptides released from brain slice preparations.[126] Effects of peptidase inhibitors on electrophysiological recording from single neurons may also prove a useful approach. Colocalization of peptides and peptidase can also provide strong supporting evidence for a functional role.[124] These types of experiments require the use of potent and selective peptidase inhibitors as well as a detailed knowledge of the active site and catalytic mechanism of individual peptidases.

V. Design of Peptidase Inhibitors

A. INHIBITION OF NEUROTRANSMITTER METABOLISM

Inhibitors of neurotransmitter metabolizing enzymes have made a substantial contribution to our understanding of synaptic biochemistry and, in some cases, have proved to be of therapeutic value. Particular examples include the use of acetylcholinesterase inhibitors in the treatment of myasthenia gravis, monoamine oxidase inhibition in depression and GABA-transaminase inhibitors as anti-convulsants. It is not surprising, therefore, that the design of inhibitors of neuropeptide metabolizing enzymes has come into prominence in recent years with the focus of attention being "enkephalinase". Any therapeutic applications have so far failed to materialize, partly as a result of the difficulty of obtaining orally active peptide analogues that are CNS accessible. A second problem, discussed above, is that of the lack of selectivity of the peptidases. Thus, inhibitors of "enkephalinase" might also be expected to modify the metabolism, and hence the actions, of substance P and perhaps other neuropeptides.

The major cell-surface peptidases implicated in neuropeptide metabolism—endopeptidase-24.11, ACE and aminopeptidases—are all zinc metallopeptidases showing common mechanistic aspects with each other and with two well characterized enzymes, carboxypeptidase A and thermolysin. Since these enzymes have proved to be the starting point for inhibitor design a brief comparison of their properties and inhibition is relevant.

B. CARBOXYPEPTIDASE A AND ANGIOTENSIN CONVERTING ENZYME

At first sight carboxypeptidase A (CP-A) has little in common with ACE or endopeptidase-24.11 in terms of localization, regulation or function. It is an extracellular enzyme synthesized and secreted by the pancreas which is involved in peptide hydrolysis in the small intestine. The enzyme is monomeric (307 amino acid residues) of M_r 35 000 and contains no bound carbohydrate. The common features among CP-A and ACE are the inclusion of a single Zn^{2+} ion per mole and the ability to remove sequentially residues from the C-terminus of a peptide, one at a time in the case of CP-A and two at a time by ACE.

It was the detailed enzymological studies of Lipscomb and collaborators[138] in the 1960s that led to an understanding of the catalytic mechanism of CP-A. There appear to be three residues critical for catalysis: Arg-145, Tyr-248 and Glu-270. Active site modification studies have likewise implicated arginyl, tyrosyl and glutamyl residues at the active site of ACE. In the case of CP-A, the Arg-145 residue at the active site forms an ionic bond with the terminal carboxyl group of the substrate. An adjacent hydrophobic pocket is responsible for its specificity towards substrates containing a C-terminal aromatic amino acid. Enzyme–substrate complexes are also characterized by interaction between the active site zinc and the carbonyl oxygen of the scissile peptide bond. Corresponding complexes between enzyme and inhibitor can produce potent inhibition. The key inhibitors of CP-A that have led to the design of ACE inhibitors were D-benzyl succinate and various thiol inhibitors. By analogy with the inhibition of CP-A, Ondetti and colleagues[139] produced the first of the orally active ACE inhibitors, the thiol compound captopril (D-3-mercapto-2-propanoyl-L-proline) which is structurally analogous to the dipeptide Ala–Pro but stable to hydrolysis by peptidases. This design reflected the potency of Ala–Pro present in a number of snake venom peptides that are potent inhibitors of ACE. Potency was achieved by replacing the carboxyl group in venom peptides with a sulphydryl group which provided a stronger ligand to the zinc ion. The K_i for captopril is approximately 5 nM and its postulated mode of interaction with the active site of the enzyme is shown in Fig. 9. Although captopril has proved a very effective anti-hypertensive drug, it is not without side-effects which have been attributed to the SH functional group. Intensive search for new classes of ACE inhibitors *in vivo* has led to the carboxyalkyl-dipeptides of which enalapril is the prototype. This methyl ester is converted to the active inhibitor (enalaprilat) *in vivo*. The lysine analogue (lisinopril) is also a potent inhibitor and does not require de-esterification *in vivo*.[140] It is

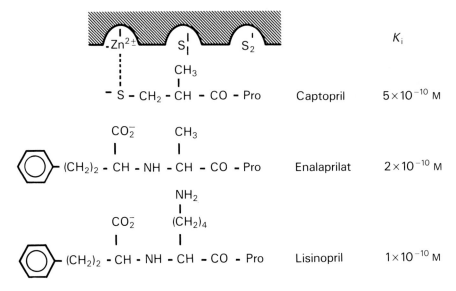

Fig. 9. Schematic diagram representing the binding of inhibitors to the active site of angiotensin converting enzyme. The three inhibitors all feature an anionic group (thiol or carboxyl) that can interact with the active-site Zn as well as a C-terminal proline. The free carboxyl group of the proline interacts with a positively charged arginyl residue (not shown) in the active site.

currently undergoing clinical trials. These inhibitors, whose structures are also given in Fig. 9, can be used for affinity purification of the enzyme and may prove useful for exploring the physiological functions of ACE other than in the metabolism of bradykinin and angiotensin I. The selectivity and potency of the current generation of inhibitors are such that they can be used, in radiolabelled form, for autoradiographic localization. [^{3}H]-captopril has particularly been used in this context and first allowed the striatonigral localization of the enzyme to be visualized.[137]

C. THERMOLYSIN AND "ENKEPHALINASE"

The strategies used for the design of CP-A and ACE inhibitors have been adapted for the development of potent and selective endopeptidase-24.11 inhibitors. A dipeptide analogue from *Streptomyces tanashiensis*, phosphoramidon, was the first potent compound to be described as an inhibitor of the enzyme[141] and was originally isolated as an inhibitor of the bacterial Zn-containing endoprotease thermolysin. The mammalian endopeptidase has certain catalytic features in common with thermolysin, particularly specificity for hydrolysis on the amino side of hydrophobic residues.[110] Of particular value to inhibitor design has been the crystallographic study of the thermolysin–phosphoramidon complex.[142] In the case of this inhibitor a variety of interactions contribute to the binding energy. These include interaction of the phosphoryl group of the inhibitor with the zinc ion, whose binding energy is potentiated by a complex H-bond network involving the active site Tyr, Glu and His residues. Furthermore the Trp of the inhibitor is in hydrophobic bonding distance of a Phe in the active site of the enzyme. As a consequence of the detailed knowledge of these interactions a variety of phosphoramides, phosphonic acids and phosphinic acids have been explored as inhibitors of endopeptidase-24.11, and indeed of ACE, and intensive research continues in this area. However, the compound that has received greatest attention as an inhibitor of enkephalin metabolism is thiorphan.[143] This compound was designed by analogy with captopril and contains a thiol group that can ligand with the Zn^{2+} and a hydrophobic amino acid (Phe) in the P_1', position (see Fig. 10). A number of ingenious variations on this structure have since been developed (see e.g. ref. 117). Reversal of the terminal amide bond led to *retro*-thiorphan, which was more selective than thiorphan and resistant to hydrolysis by carboxypeptidases.[144] Acetorphan, an esterified form of thiorphan, is several orders of magnitude less potent than thiorphan, but can gain access to the CNS where it can be hydrolysed to generate thiorphan itself. Finally, kelatorphan has been designed[145] to inhibit not only endopeptidase-24.11 but also aminopeptidase activity and, to a lesser degree, ACE. Since it therefore inhibits all of the enzymes implicated in enkephalin metabolism, it may have some value in potentiating the physiological actions of the enkephalins.

Such evidence as is available from inhibitor studies substantiates a role for endopeptidase-24.11 in terminating enkephalin action but by no means excludes a role for the enzyme in hydrolysing other biologically active peptides. Several techniques have been devised to assess the role of individual peptidases. In particular, thiorphan has been shown to protect enkephalin from metabolism when released from brain slices by depolarization with potassium. However, complete recovery of enkephalin is only achieved in the simultaneous presence of an aminopeptidase inhibitor.[125,126] Phosphoramidon can protect from hydrolysis by synaptic membranes, cholecystokinin, substance P and other tachykinins.[109,121,122,146] There are some reports of analgesic activity of enkephalinase inhibitors, but their efficiency is questionable.[117,118] Besides analgesics, these inhibitors can also elicit a variety of other opiate-like effects, some of which are prevented by opiate antagonists such as naloxone.

The development of new peptidase inhibitors has frequently shown that nature can produce compounds with a selectivity and potency to equal or better the designs of synthetic chemists. Snake venom peptide inhibitors

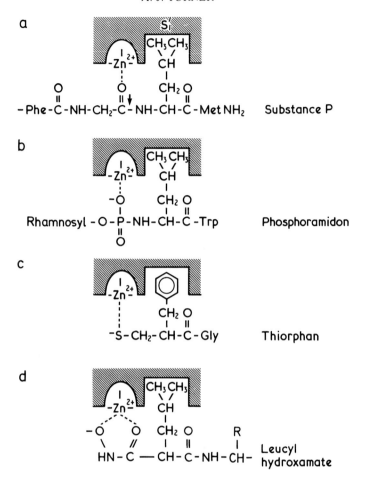

a

$-Phe-\overset{O}{\overset{\|}{C}}-NH-CH_2-\overset{O}{\overset{\|}{C}}-NH-CH-\overset{O}{\overset{\|}{C}}-Met\,NH_2$ Substance P

b

$Rhamnosyl-O-\overset{O}{\underset{\|}{\overset{|}{P}}}-NH-CH-\overset{O}{\overset{\|}{C}}-Trp$ Phosphoramidon

c

$^-S-CH_2-CH-\overset{O}{\overset{\|}{C}}-Gly$ Thiorphan

d

$HN-C-CH-\overset{O}{\overset{\|}{C}}-NH-CH-$ Leucyl hydroxamate

Fig. 10. Schematic diagram representing the binding of substrate and inhibitors to endo-peptidase-24.11. The active site features a Zn^{2+} ion and a hydrophobic site which can interact with the appropriate hydrophobic group in the P_1' position of the substrate or inhibitor. (a) The C-terminal tetrapeptide region (—Phe—Gly—Leu—MetNH$_2$) of the undecapeptide, substance P. The principal site of hydrolysis by the enzyme is the Gly^9–Leu^{10} bond. (b) The natural product inhibitor phosphoramidon (see ref. 141). (c) The synthetic thiol inhibitor, thiorphan (see ref. 142). (d) A bidentate hydroxamate inhibitor of endopeptidase-24.11 (see ref. 119).

played a crucial role in establishing the clinical value of ACE inhibitors in hypertension and the search continues for new compounds. Umezawa and colleagues[146] have been foremost in this regard. One of the most potent compounds recently isolated is a peptide fragment arising from tryptic digestion of a plasma protein fraction.[147] The active peptide, designated converstatin, is a remarkably potent competitive inhibitor of ACE with a K_i value of $4·5 \times 10^{-12}$ M with a sequence corresponding to that of bradykinin, but lacking the prolyl residue in position 3 (i.e. [des-Pro^3]-bradykinin). It is unclear why deletion of the third amino acid residue of bradykinin should produce such a dramatic increase in binding affinity and emphasizes that the binding of peptide substrates or inhibitors to peptidases involves interactions of residues distant from the bond to be hydrolysed. The search for natural inhibitors of endopeptidase-24.11 has been less extensive but the discovery of phosphoramidon as an inhibitor of the enzyme[141] has been the key to the development of many new inhibitors. Their exploitation for probing the active site, localization and physiological functions of these enzymes should follow in the next few years.

VI. Likely Future Developments

Over the last decade it has become apparent that peptidergic neurons are a major component of the mammalian nervous system. Recent advances in molecular biology have contributed to a large degree in these developments and have provided new insights into the regulation of eukaryotic gene expression and intercellular communication. Neuropeptides can occur as multiple precursors within a single gene product or as closely related gene families whose end-products can be regulated at transcriptional and translational levels on a tissue-specific basis. The termination of peptide signals appears to occur predominantly through the actions of a discrete set of cell-surface peptidases that are neither peptide- nor tissue-specific.

The full diversity of actions of neuropeptides has yet to be appreciated. Just as noradrenaline can function both as neurotransmitter and as a hormone, so too can the peptides exert distinct effects on different target tissues. In addition to their roles in neural communication and in stress situations, opioid peptides appear to be able to modulate the immune system.[148,149] For example, lymphocytes possess specific binding sites for β-endorphin that are distinct from classical opiate receptors. In this case it is the C-terminal portion of the peptide (the non-opioid fragment) that is recognized by the receptor. Natural opioid peptides appear to regulate the maturation or activation of natural killer (NK) cells, proliferation of T-lymphocytes and also the growth of certain tumours. Remarkably, β-endorphin shows some actions and also some antigenic cross-reactivity with human leukocyte interferon. Other neuropeptides can also regulate cell division. Both substance P and neurokinin A can act as mitogens for connective tissue and this effect appears to be mediated through a particular subclass of tachykinin receptor.[150] The peptide vasopressin also shows mitogenic activity.[151] Studies of changes in neuropeptide systems in neurological and psychological disorders are, as yet, in their early stages. Decreases in the levels of substance P, enkephalins, CCK and angiotensin have been detected in post-mortem brains of patients suffering from Huntington's Chorea. The levels of somatostatin and CRF appear to be selectively reduced in the cerebral cortex in Alzheimer's disease[152] (pre-senile dementia). Changes in peptide levels in schizophrenic brains have also been noted. Whether these changes are primary or secondary deficits is unclear, and their functional significance to the disease states will require much further evaluation.

The advances of the last few years bear remarkable testimony to the combined efforts of biochemists, pharmacologists, neuroanatomists and molecular biologists. We shall no doubt see the appearance of new peptide candidates whose distribution will be mapped. The emphasis of the molecular biologists will move towards the factors regulating the expression of neuropeptide genes and the processing enzymes will, in due course, be isolated and characterized. Characterization and isolation of neuropeptide receptors and the development of suitable antagonists will also be a priority. The goal of therapeutic agents that modify neuropeptide systems is still some way off and factors such as selectivity of action and accessibility to the CNS have yet to be overcome. It is to be hoped that the next decade will succeed in unifying our concepts regarding the multiplicity of regulatory peptides, their biochemical actions and their diverse cellular targets.

ACKNOWLEDGEMENTS

Aspects of the research reported here have been supported by the Medical Research Council; the Wellcome Trust; Smith, Kline Foundation; Merck, Sharp & Dohme Research Laboratories and the University of Leeds Research Fund. I should like to thank Dr John Kenny for introducing me to the world of peptidases.

REFERENCES

1. Krieger, D. T. (1983). Brain peptides: what, where, and why? *Science (N.Y.)* **222**, 975–985.
2. Lynch, D. R. & Snyder, S. H. (1986). Neuropeptides: multiple molecular forms, metabolic pathways and receptors. *Annu. Rev. Biochem.* **55**, 773–779.
3. Pickering, B. T. (1978). The neurosecretory neurone: a model system for the study of secretion. In *Essays in Biochemistry* Vol. 14 (Campbell, P. N. & Aldridge, W. N., eds) pp. 45–81. Academic Press, London.
4. Barker, J. L. (1976). Peptides: roles in neuronal excitability. *Physiol. Rev.* **56**, 435–452.
5. Hughes, J., Smith, T., Morgan, B. & Fothergill, L. (1975). Purification and

properties of enkephalin—the possible endogenous ligand for the morphine receptor. *Life Sci.* **16**, 1753–1758.

6. Simantov, R. & Snyder, S. H. (1976). Morphine-like peptides in mammalian brain: isolation, structure elucidation and interactions with the opiate receptor. *Proc. Natl. Acad. Sci. USA* **73**, 2515–2519.

7. Roisin, M. P., Artola, A., Henry, J. P. & Rossier, J. (1983). Enkephalins are associated with adrenergic granules in bovine adrenal medulla. *Neuroscience* **10**, 83–88.

8. Flynn, T. G., de Bold, M. L. & de Bold, A. J. (1983). The amino acid sequence of an atrial peptide with potent diuretic and natriuretic properties. *Biochem. Biophys. Res. Commun.* **117**, 859–865.

9. Currie, M. G., Geller, D. M., Cole, B. R., Siegel, N. R., Fok, K. F., Adams, S. P., Eubanks, S. R., Gallupi, G. R. & Needleman, P. (1984). Purification and sequence analysis of bioactive atrial peptides (atriopeptins). *Science* **223**, 67–69.

10. Jacobowitz, D. M., Skofitsch, G., Keiser, H. R., Eskay, R. L. & Zamir, N. (1985). Evidence for the existence of atrial natriuretic factor-containing neurons in the rat brain. *Neuroendocrinology* **40**, 92–94.

11. Carraway, R., Leeman, S. E. & Niall, H. D. (1971). The isolation of a new hypotensive peptide, neurotensin, from bovine hypothalamus. *J. Biol. Chem.* **248**, 6854–6861.

12. Hanley, M. R. (1985). Neuropeptides as mitogens. *Nature (London)* **315**, 14–15.

13. Chang, K.-J. (1984). Opioid peptides have actions on the immune system. *Trends Neurosci.* **7**, 234–235.

14. White, J. D., Stewart, K. D., Krause, J. E. & McKelvy, J. F. (1985). Biochemistry of peptide-secreting neurons. *Physiol. Rev.* **65**, 553–606.

15. Rosenfeld, M. G., Mermod, J. J., Amara, S. G., Swanson, L. W., Sawchenko, P. E., Rivier, J., Vale, W. W. & Evans, R. H. (1983). Production of a novel neuropeptide encoded by the calcitonin gene via tissue-specific RNA processing. *Nature (London)* **304**, 129–135.

16. Tatemoto, K. (1982). Neuropeptide Y: complete amino acid sequence of the brain peptide. *Proc. Natl. Acad. Sci. USA* **79**, 5485–5489.

17. Sutcliffe, J. G. & Milner, R. J. (1984). Brain specific gene expression. *Trends Biochem. Sci.* **9**, 95–99.

18. Palkovits, M. (1984). Distribution of neuropeptides in the central nervous system: a review of biochemical mapping studies. *Progr. Neurobiol.* **23**, 151–189.

19. Vanderhaegen, J.-J., Signeau, J. C. & Gepts, W. (1975). New peptide in the vertebrate CNS reacting with gastrin antibodies. *Nature (London)* **257**, 604–605.

20. Gray, T. S. & Morley, J. E. (1986). Neuropeptide Y: anatomical distribution and possible function in mammalian nervous system. *Life Sci.* **38**, 389–401.

21. Coghlan, J. P., Aldred, P., Haralambidis, J., Niall, H. D., Penschow, J. D. & Tregear, G. W. (1985). Hybridization histochemistry. *Anal. Biochem.* **149**, 1–28.

22. Gee, C. E., Chen, C. L. C., Roberts, J. L., Thompson, R. & Watson, S. J. (1983). Identification of pro-opiomelanocortin neurones in rat hypothalamus by *in situ* cDNA–mRNA hybridization. *Nature (London)* **306**, 374–376.

23. Hokfelt, T., Johansson, O., Ljundahl, A., Lundberg, J. M. & Schultzberg, M. (1980). Peptidergic neurones. *Nature (London)* **284**, 515–521.
24. Johansson, O. & Lundberg, J. M. (1981). Ultrastructural localisation of VIP-like immunoreactivity in large dense-core vesicles of "cholinergic type" nerve terminals in cat exocrine glands. *Neuroscience* **6**, 847–862.
25. O'Donohue, T. L., Millington, W. R., Handelmann, G. E., Contreras, P. C. & Chronwall, B. M. (1985). On the 50th Anniversary of Dale's Law: multiple neurotransmitter neurons. *Trends Pharmacol. Sci.* **6**, 305–308.
26. Hendry, S. H. C., Jones, E. G., DeFelipe, J., Schmechel, D., Brandon, C. & Emson, P. C. (1984). Neuropeptide-containing neurons of the cerebral cortex are also GABAergic. *Proc. Natl. Acad. Sci. USA* **81**, 6526–6530.
27. Iversen, L. L., Iversen, S. D., Bloom, F., Douglas, C., Brown, M. & Vale, W. (1978). Calcium-dependent release of somatostatin and neurotensin from rat brain *in vitro. Nature (London)* **273**, 161–163.
28. Osborne, N. N. (ed.) (1983). *Dale's Principle and Communication between Neurons.* Pergamon Press, Oxford.
29. Udenfriend, S. & Kilpatrick, D. L. (1983). Biochemistry of the enkephalins and enkephalin-containing peptides. *Arch. Biochem. Biophys.* **221**, 309–323.
30. Akil, H., Watson, S. J., Young, E., Lewis, M. E., Khachaturian, H. & Walker, J. M. (1984). Endogenous opioids: biology and function. *Annu. Rev. Neurosci.* **7**, 223–255.
31. Hughes, J., Smith, T. W., Kosterlitz, H. W., Fothergill, L. H., Morgan, B. A. & Morris, H. R. (1975). Identification of two related pentapeptides from the brain with potent opiate agonist activity. *Nature (London)* **258**, 577–579.
32. Bradbury, A. F., Smyth, D. G., Snell, C. R., Hulme, E. C. & Birdsall, N. J. M. (1976). C-fragment of lipotropin has a high affinity for opiate receptors. *Nature (London)* **260**, 793–795.
33. Zakarian, S. & Smyth, D. G. (1982). Distribution of β-endorphin related peptides in rat pituitary and brain. *Biochem. J.* **202**, 561–571.
34. Douglass, J., Civelli, O. & Herbert, E. (1984). Polyprotein gene expression: generation of diversity of neuroendocrine peptides. *Annu. Rev. Biochem.* **53**, 665–715.
35. Horikawa, S., Takai, T., Toyosato, M., Takahashi, H., Noda, M., Kakidani, H., Kubo, T., Hirose, T., Inayama, S., Hayashida, H., Miyata, M. & Numa, S. (1983). Isolation and structural organization of the human pro-enkephalin B gene. *Nature (London)* **306**, 611–614.
36. Noda, M., Furutani, Y., Takahashi, H., Toyosato, M., Hirose, T., Inayama, S., Nakanishi, S. & Numa, S. (1982). Cloning and sequence analysis of cDNA for bovine adrenal preproenkephalin. *Nature (London)* **295**, 202–206.
37. Gubler, U., Seeburg, P., Hoffman, B. J., Gage, L. P. & Udenfriend, S. (1982). Molecular cloning establishes proenkephalin as precursor of enkephalin-containing peptides. *Nature (London)* **295**, 206–208.
38. Comb, M., Seeburg, P. H., Adelman, J., Eiden, L. & Herbert, E. (1982). Primary structure of the human Met- and Leu-enkephalin precursor and its mRNA. *Nature (London)* **295**, 663–666.
39. Kakidani, H., Furutani, Y., Takahashi, H., Noda, M., Morimoto, Y., Hirose, T., Asai, M., Inayama, S., Nakanishi, S. & Numa, S. (1982). Cloning and sequence analysis of cDNA for porcine β-neo-endophin/dynorphin precursor. *Nature (London)* **298**, 245–249.

40. Goldstein, A., Fischli, W., Lowney, L. I., Hunkapiller, M. & Hood, L. (1981). Bovine pituitary dynorphin: complete amino acid sequence of the biologically active heptadecapeptide. *Proc. Natl. Acad. Sci. USA* **78**, 7219–7223.
41. Liston, D., Patey, G., Rossier, J., Verbanck, P. & Vanderhaegen, J.-J. (1984). Processing of proenkephalin is tissue-specific. *Science* **225**, 734–737.
42. Von Euler, U. S. & Gaddum, J. H. (1931). An unidentified depressor substance in certain tissue extracts. *J. Physiol.* **72**, 74–87.
43. Von Euler, U. S. & Pernow, B. (eds) (1977). In *Substance P*. Raven Press, New York.
44. Chang, M. M., Leeman, S. E. & Niall, H. D. (1971). Amino acid sequence of substance P. *Nature New Biol.* **232**, 86–87.
45. Otsuka, M. & Konishi, S. (1976). Release of substance P-like immunoreactivity from isolated spinal cord of newborn rat. *Nature* (*London*) **264**, 83–84.
46. Minamino, N., Masuda, H., Kangawa, K. & Matsuo, H. (1984). Neuromedin L: a novel mammalian tachykinin identified in porcine spinal cord. *Neuropeptides* **4**, 157–166.
47. Nawa, H., Doteuchi, M., Igano, K., Inouye, K. & Nakanishi, S. (1984). Substance K: a novel mammalian tachykinin that differs from substance P in its pharmacological profile. *Life Sci.* **34**, 1153–1160.
48. Maggio, J. E. & Hunter, J. C. (1984). Regional distribution of kassinin-like immunoreactivity in rat central and peripheral tissues and the effect of capsaicin. *Brain. Res.* **307**, 370–373.
49. Nawa, T., Hirose, T., Takashima, H., Inayama, S. & Nakanishi, S. (1983). Nucleotide sequences of cloned cDNAs for two types of bovine brain substance P precursor. *Nature* (*London*) **306**, 32–36.
50. Nawa, H., Kotani, H. & Nakanishi, S. (1984). Tissue specific generation of two preprotachykinin mRNAs from one gene by alternative RNA splicing. *Nature* (*London*) **312**, 729–734.
51. Buck, S. H., Burcher, E., Shults, C. W., Lovenberg, W. & O'Donohue, T. L. (1984). Novel pharmacology of substance K-binding sites: a third type of tachykinin receptor. *Science* (*N.Y.*) **226**, 987–989.
52. Craig, R. H. (1982). Partial nucleotide sequence of human calcitonin precursor mRNA identifies flanking cryptic peptides. *Nature* (*London*) **295**, 345–347.
53. Amara, S. G., Arriza, J. L., Leff, S. E., Swanson, L. W., Evans, R. M. & Rosenfeld, M. G. (1985). Expression in brain of a messenger RNA encoding a novel neuropeptide homologous to calcitonin gene-related peptide. *Science* **229**, 1094–1097.
54. Fisher, L. A., Kikkawa, D. O., Rivier, J. E., Amara, S. G., Evans, R. M., Rosenfeld, M. G., Vale, W. W. & Brown, M. R. (1983). Stimulation of noradrenergic sympathetic outflow by calcitonin gene-related peptide. *Nature* (*London*) **305**, 534–536.
55. Nemer, N., Chamberland, M., Sirois, D., Argentin, S., Drouin, J., Dixon, R. A. F., Zivin, R. A. & Condra, J. W. (1984). Gene structure of human cardiac hormone precursor, pronatriodilatin. *Nature* (*London*) **312**, 654–656.
56. Jamieson, J. D. & Palade, G. E. (1964). Specific granules in atrial muscle cells. *J. Cell Biol.* **23**, 151–172.
57. de Bold, A. J. (1985). Atrial natriuretic factor: a hormone produced by the heart. *Science* (*N.Y.*) **230**, 767–770.
58. Flynn, T. G. & Davies, P. L. (1985). The biochemistry and molecular biology of atrial natriuretic factor. *Biochem. J.* **232**, 313–321.

59. Turner, A. J. (1985). Hormones from the heart. *Trends Biochem. Sci.* **10**, 2–3.
60. Zamir, N., Skofitsch, G., Eskay, R. L. & Jacobowitz, D. M. (1986). Distribution of immunoreactive atrial natriuretic peptides in the central nervous system of the rat. *Brain Res.* **365**, 105–111.
61. Iversen, L. L. (1983). Non-opioid neuropeptides in mammalian CNS. *Annu. Rev. Pharmacol. Toxicol.* **23**, 1–27.
62. Blobel, G. & Dobberstein, B. (1975). Transfer of proteins across membranes: I Presence of proteolytically processed and unprocessed nascent immunoglobulin light chains on membrane-bound ribosomes of murine myeloma. *J. Cell Biol.* **67**, 835–851.
63. Evans, E. A., Gilmore, R. & Blobel, G. (1986). Purification of microsomal signal peptidase as a complex. *Proc. Natl. Acad. Sci. USA* **83**, 581–585.
64. Loh, Y. P., Brownstein, M. J. & Gainer, H. (1984). Proteolysis in neuropeptide processing and other neural functions. *Annu. Rev. Neurosci.* **7**, 189–222.
65. Turner, A. J. (1984). Neuropeptide processing enzymes. *Trends Neurosci.* **7**, 258–260.
66. Kreil, G., Haiml, L. & Suchanek, G. (1980). Stepwise cleavage of the pro part of promelittin by dipeptidyl peptidase IV: evidence for a new type of precursor-product conversion. *Eur. J. Biochem.* **111**, 49–58.
67. Docherty, K. & Steiner, D. F. (1982). Post-translational proteolysis in polypeptide hormone biosynthesis. *Annu. Rev. Physiol.* **44**, 625–638.
68. Roberts, J. L. & Pritchett, D. (1984). Does the Kallikrein-like enzyme gene family code for a group of peptide hormone-processing enzymes? *Trends Neurosci.* **7**, 105–107.
69. Berger, E. A. & Shooter, E. M. (1977). Evidence for pro-β-nerve growth factor, a biosynthetic precursor to β-nerve growth factor. *Proc. Natl. Acad. Sci. USA* **74**, 3647–3651.
70. Frey, P., Forand, R., Maciag, T. & Shooter, E. M. (1979). The biosynthetic precursor of epidermal growth factor and the mechanism of its processing. *Proc. Natl. Acad. Sci. USA* **76**, 6294–6298.
71. Owen, M. C., Brennan, S. O., Lewis, J. H. & Carrell, R. W. (1983). Mutation of anti-trypsin to anti-thrombin: α_1-antitrypsin Pittsburgh (358 Met $\rightarrow$ Arg), a fatal bleeding disorder. *New Engl. J. Med.* **309**, 694–698.
72. Brennan, S. O. & Carrell, R. W. (1978). A circulating variant of human proalbumin. *Nature (London)* **274**, 908–909.
73. Green, D. P. L. (1985). Are prohormone activating enzymes part of an enzyme cascade? *Trends Pharmacol. Sci.* **8**, 11–12.
74. Docherty, K., Carroll, R. & Steiner, D. F. (1983). Identification of a 31 000 molecular weight islet cell protease as cathepsin B. *Proc. Natl. Acad. Sci. USA* **80**, 3245–3249.
75. Loh, Y. P. & Gainer, H. (1982). Characterisation of pro-opiocortin-converting activity in purified secretory granules from rat pituitary neurointermediate lobe. *Proc. Natl. Acad. Sci. USA* **79**, 108–112.
76. Troy, C. M. & Musacchio, J. M. (1982). Processing of enkephalin precursors by chromaffin granule membranes. *Life Sci.* **31**, 1717–1720.
77. Moore, H.-P., Gumbiner, B. & Kelly, R. B. (1983). A subclass of proteins and sulfated macromolecules secreted by AtT-20 (mouse pituitary tumor) cells is sorted with adrenocorticotropin into dense secretory granules. *J. Cell Biol.* **97**, 810–817.
78. Julius, D., Brake, A., Blair, L., Kunisawa, R. & Thorner, J. (1984). Isolation

of the putative structural gene for the lysine-arginine-cleaving endopeptidase required for processing of yeast pre-pro-α-factor. *Cell* **37**, 1075–1089.

79. Schwartz, T. W. (1986). "Proline directed arginyl cleavage" and other monobasic processing mechanisms in peptide biogenesis. *FEBS Lett.* **200**, 1–10.

80. Eng, J., Shiina, Y., Pan, Y. C. E., Blacher, R., Chang, M., Stein, S. & Yalow, R. S. (1983). Pig brain contains cholecystokinin octapeptide and several cholecystokinin desoctapeptides. *Proc. Natl. Acad. Sci. USA* **80**, 6381–6385.

81. Hook, V. Y. H., Eiden, L. E. & Brownstein, M. J. (1982). A carboxypeptidase processing enzyme for enkephalin precursors. *Nature (London)* **295**, 341–342.

82. Supattapone, S., Fricker, L. D. & Snyder, S. H. (1984). Purification and characterisation of a membrane-bound enkephalin-forming carboxypeptidase, "enkephalin convertase". *J. Neurochem.* **42**, 1017–1023.

83. Fricker, L. D. & Snyder, S. H. (1982). Enkephalin convertase: Purification and characterisation of a specific enkephalin synthesizing carboxypeptidase localized to adrenal chromaffin granules. *Proc. Natl. Acad. Sci. USA* **79**, 3886–3890.

84. Fricker, L. D. (1985). Neuropeptide biosynthesis: focus on the Carboxypeptidase processing enzyme. *Trends Neurosci.* **8**, 210–214.

85. Fricker, L. D. & Snyder, S. H. (1983). Purification and characterization of enkephalin convertase, an enkephalin synthesizing carboxypeptidase. *J. Biol. Chem.* **258**, 10950–10955.

86. Fricker, L. D., Plummer, T. H. & Snyder, S. H. (1983). Enkephalin convertase: potent, selective and irreversible inhibitors. *Biochem. Biophys. Res. Commun.* **111**, 994–1000.

87. Lynch, D. R., Strittmatter, S. M. & Snyder, S. H. (1984). Enkephalin convertase localization by (^{3}H)guanidinoethylmercaptosuccinic acid antoradiography: selective association with enkephalin-containing neurons. *Proc. Natl. Acad. Sci. USA* **81**, 6543–6547.

88. Bradbury, A. F., Finnie, M. D. A. & Smyth, D. G. (1982). Mechanism of C-terminal amide formation by pituitary enzymes. *Nature (London)* **298**, 686–688.

89. Hilsted, L., Rehfeld, J. F. & Schwartz, T. W. (1986). Impaired α-carboxy-amidation of gastrin in vitamin C-deficient guinea pigs. *FEBS Lett.* **196**, 151–154.

90. May, V. & Eipper, B. A. (1985). Regulation of peptide amidation in cultured pituitary cells. *J. Biol. Chem.* **260**, 16224–16231.

91. O'Donohue, T. L. (1983). Identification of endorphin acetyltransferase in rat brain and pituitary gland. *J. Biol. Chem.* **258**, 2163–2167.

92. Smyth, D. G., Massey, D. E., Zakarian, S. & Finnie, M. D. (1979). Endorphins are stored in biologically active and inactive forms: isolation of α-N-acetyl peptides. *Nature (London)* **279**, 252–254.

93. Lee, R. W. H. & Huttner, W. B. (1985). (Glu^{62}, Ala^{30}, Tyr^8)$_n$ serves as high-affinity substrate for tyrosylprotein sulfotransferase: A Golgi enzyme. *Proc. Natl. Acad. Sci. USA* **82**, 6143–6147.

94. Uhler, M. & Herbert, E. (1983). Complete amino acid sequence of mouse pro-opiomelanocortin derived from the nucleotide sequence of pro-opiomelanocortin cDNA. *J. Biol. Chem.* **258**, 257–261.

95. Rehfeld, J. F. (1985). Neuronal cholecystokinin: one or multiple transmitters. *J. Neurochem.* **44**, 1–10.
96. Liston, D., Patey, G., Rossier, J., Verbanck, P. & Vanderhaegen, J.-J. (1984). Processing of proenkephalin is tissue-specific. *Science* (*N.Y.*) **225**, 734–737.
97. Vale, W., Speiss, J., Rivier, C. & Rivier, J. (1981). Characterization of a 41-residue ovine hypothalamic peptide that stimulates secretion of corticotropin and β-endorphin. *Science* **213**, 1394–1397.
98. Chen, C. L. C., Dionne, F. T. & Roberts, J. L. (1983). Regulation of the pro-opiomelanocortin mRNA levels in rat pituitary by dopaminergic compounds. *Proc. Natl. Acad. Sci. USA* **80**, 2211–2215.
99. Tang, F., Costa, E. & Schwartz, J. P. (1983). Increase of proenkephalin mRNA and enkephalin content of rat striatum after daily injection of haloperidol for 2 to 3 weeks. *Proc. Natl. Acad. Sci. USA* **80**, 3841–3844.
100. Smyth, D. G. (1985). Regulation of processing in multi-functional prohormones: a new hypothesis. *Biochem. Soc. Trans.* **13**, 38–39.
101. Turner, A. J. & Whittle, S. R. (1983). Biochemical dissection of the γ-aminobutyrate synapse. *Biochem. J.* **209**, 29–41.
102. Radian, R. & Kanner, B. I. (1985). Reconstitution and purification of the sodium and chloride-coupled γ-aminobutyric acid transporter from rat brain. *J. Biol. Chem.* **260**, 11 859–11 865.
103. Hambrook, J. M., Morgan, B. A., Rance, M. J. & Smith, C. F. C. (1976). Mode of deactivation of the enkephalins by rat and human plasma and rat brain homogenates. *Nature* (*London*) **262**, 782–783.
104. Meek, J. L., Yang, H. Y. T. & Costa, E. (1977). Enkephalin catabolism *in vitro* and *in vivo*. *Neuropharmacology* **16**, 151–154.
105. Schwartz, J.-C., Malfroy, B. & de la Baume, S. (1981). Biological inactivation of enkephalins and the role of enkephalin-dipeptidylcarboxypeptidase ("enkephalinase") as neuropeptidase. *Life Sci.* **29**, 1715–1740.
106. Schwartz, J.-C. (1983). Metabolism of enkephalins and the inactivating neuropeptidase concept. *Trends Neurosci.* **6**, 45–48.
107. Lee, C. M., Sandberg, B. E. B., Hanley, M. R. & Iversen, L. L. (1981). Purification and characterisation of a membrane-bound substance P degrading enzyme from human brain. *Eur. J. Biochem.* **114**, 315–327.
108. Fulcher, I. S., Matsas, R., Turner, A. J. & Kenny, A. J. (1982). Kidney neutral endopeptidase and the hydrolysis of enkephalin by synaptic membranes show similar sensitivity to inhibitors. *Biochem. J.* **203**, 519–522.
109. Matsas, R., Fulcher, I. S., Kenny, A. J. & Turner, A. J. (1983). Substance P and (Leu) enkephalin are hydrolyzed by an enzyme in pig caudate synaptic membranes that is identical with the endopeptidase of kidney microvilli. *Proc. Natl. Acad. Sci. USA* **80**, 3111–3115.
110. Turner, A. J., Matsas, R. & Kenny, A. J. (1985). Are there neuropeptide-specific peptidases? *Biochem. Pharmacol.* **34**, 1347–1356.
111. Kenny, A. J. & Booth, A. G. (1978). Microvilli: their ultrastructure, enzymology and molecular organization. In *Essays in Biochemistry*, vol. 14, pp. 1–44. Academic Press, London.
112. Kenny, A. J. & Maroux, S. (1982). Topology of microvillar membrane hydrolases of kidney and intestine. *Physiol. Rev.* **62**, 91–128.
113. Relton, J. M., Gee, N. S., Matsas, R., Turner, A. J. & Kenny, A. J. (1983). Purification of endopeptidase-24.11 ('enkephalinase') from pig brain by immunoadsorbent chromatography. *Biochem. J.* **215**, 519–523.

114. Matsas, R., Stephenson, S. L., Hryszko, J., Kenny, A. J. & Turner, A. J. (1985). The metabolism of neuropeptides: Phase separation of synaptic membrane preparations with Triton X-114 reveals the presence of aminopeptidase N. *Biochem. J.* **231**, 445–449.

115. Graves, F. B., Law, P. Y., Hunt, C. A. & Loh, H. H. (1978). The metabolic disposition of radio-labelled enkephalins *in vitro* and *in situ*. *J. Pharmacol. Exp. Therap.* **206**, 492–506.

116. Malfroy, B., Swerts, J. P., Guyon, A., Roques, B. P. & Schwartz, J. C. (1978). High-affinity enkephalin degrading peptidase is increased after morphine. *Nature (London)* **276**, 523–526.

117. Schwartz, J.-C., Costentin, J. & Lecomte, J.-M. (1985). Pharmacology of enkephalinase inhibitors. *Trends Pharmacol. Sci.* **6**, 472–476.

118. Turner, A. J., Kenny, A. J. & Matsas, R. (1986). Pharmacology of enkephalinase inhibitors. *Trends Pharmacol. Sci.* **7**, 88–89.

119. Blumberg, S., Vogel, Z. & Altstein, M. (1981). Inhibition of enkephalin-degrading enzymes from rat brain and of thermolysin by aminoacid hydroxamates. *Life Sci.* **28**, 301–306.

120. Almenoff, J. & Orlowski, M. (1984). Biochemical and immunological properties of a membrane-bound brain metalloendopeptidase: comparison with thermolysin-like kidney neutral metalloendopeptidase. *J. Neurochem.* **42**, 151–157.

121. Hooper, N. M., Kenny, A. J. & Turner, A. J. (1985). The metabolism of neuropeptides: neurokinin A (substance K) is a substrate for endopeptidase-24.11. but not peptidyl dipeptidase A (angiotensin converting enzyme). *Biochem. J.* **231**, 357–361.

122. Hooper, N. M. & Turner, A. J. (1985). Neurokinin B is hydrolysed by synaptic membranes and by endopeptidase-24.11. ('enkephalinase') but not by angiotensin converting enzyme. *FEBS Lett.* **190**, 133–136.

123. Matsas, R., Kenny, A. J. & Turner, A. J. (1984). The metabolism of neuropeptides: the hydrolysis of peptides, including enkephalins, tachykinins and their analogues by endopeptidase-24.11. *Biochem. J.* **223**, 433–440.

124. Matsas, R., Kenny, A. J. & Turner, A. J. (1986). An immunohistochemical study of endopeptidase-24.11. ('enkephalinase') in the pig nervous system. *Neuroscience* **18**, 991–1012.

125. de la Baume, S., Gros, C., Yi, C.-C., Chaillet, P., Marcais-Collado, W., Costentin, J. & Schwartz, J.-C. (1982). Selective participation of both "enkephalinase" and aminopeptidase activities in the metabolism of endogenous enkephalins. *Life Sci.* **31**, 1753–1756.

126. Chaillet, P., Marcais-Collado, W., Costentin, J., Yi, C.-C., de la Baume, S. & Schwartz, J. C. (1983). Inhibition of enkephalin metabolism by, and antinociceptive activity of, bestatin, an aminopeptidase inhibitor. *Eur. J. Pharmacol.* **86**, 329–336.

127. Gros, C., Giros, B. & Schwartz, J.-C. (1985). Identification of amino–peptidase M as an enkephalin-inactivating enzyme in rat cerebral membranes. *Biochemistry* **24**, 2179–2185.

128. Patchett, A. A. & Cordes, E. H. (1985). The design and and properties of N-carboxyalkyldipeptide inhibitors of Angiotensin converting enzyme. *Adv. Enzymol.* **57**, 1–84.

129. Soffer, R. L. (1976). Angiotensin converting enzyme and the regulation of vasoactive peptides. *Annu. Rev. Biochem.* **45**, 73–94.

130. Ganten, D., Lang, R. E., Lehmann, E. & Unger, T. (1984). Brain Angiotensin: on the way to becoming a well-studied neuropeptide system. *Biochem. Pharmacol.* **33**, 3523–3528.

131. Norman, J. A., Autry, W. L. & Barbaz, B. S. (1985). Angiotensin-converting enzyme inhibitors potentiate the analgesic activity of (Met)-enkephalin-Arg^6-Phe^7 by inhibiting its degradation in mouse brain. *Mol. Pharmacol.* **28**, 521–526.

132. Checler, F., Emson, P. C., Vincent, J. P. & Kitabgi, P. (1984). Inactivation of neurotensin by rat brain synaptic membranes. Cleavage at the Pro^{10}-Tyr^{11} bond by endopeptidase-24.11 (enkephalinase) and a peptidase different from proline-endopeptidase. *J. Neurochem.* **43**, 1295–1301.

133. Yokosawa, H., Endo, S., Ogura, Y. & Ishii, S.-I. (1983). A new feature of angiotensin-converting enzyme in the brain. Hydrolysis of substance P. *Biochem. Biophys. Res. Commun.* **116**, 735–742.

134. Cascieri, M. A., Bull, H. G., Mumford, R. A., Patchett, A. A., Thornberry, N. A. & Liang, T. (1984). Carboxy-terminal tripeptidyl hydrolysis of substance P by purified rabbit lung angiotensin converting enzyme and the potentiation of substance P activity *in vivo* by captopril and MK-422. *Mol. Pharmacol.* **25**, 287–293.

135. Skidgel, R. A., Engelbrecht, S., Johnson, A. R. & Erdös, E. G. (1984). Hydrolysis of substance P and neurotensin by converting enzyme and neutral endopeptidase. *Peptides* **5**, 769–776.

136. Skidgel, R. A. & Erdös, E. G. (1985). Novel activity of angiotensin I converting enzyme: release of the NH$_2$ and COOH-terminal tripeptides from the luteinising hormone-releasing hormone. *Proc. Natl. Acad. Sci. USA* **82**, 1025–1029.

137. Strittmatter, S. M., Lo, M. M. S., Javitch, J. A. & Snyder, S. H. (1984). Antoradiographic visualization of angiotensin converting enzyme in rat brain with (^{3}H) captopril: localization to a striatonigral pathway. *Proc. Natl. Acad. Sci. USA* **81**, 1599–1603.

138. Lipscomb, W. N. (1983). Structure and catalysis of enzymes. *Annu. Rev. Biochem.* **52**, 17–34.

139. Ondetti, M. A., Rubin, B. & Cushman, D. W. (1977). Design of specific inhibitors of angiotensin-converting enzyme: new class of orally active antihypertensive drugs. *Science (N.Y.)* **196**, 441–444.

140. Bull, H. G., Thornberry, N. A., Cordes, M. H. J., Patchett, A. A. & Cordes, E. H. (1985). Inhibition of rabbit lung angiotensin-converting enzyme by N-(S(-carboxy-3-phenylpropyl) L-alanyl-L-proline and N-((S)-1-carboxy-3-phenylproply)-L-lysyl-L-proline. *J. Biol. Chem.* **260**, 2952–2962.

141. Kenny, A. J. (1977). Proteinases associated with cell membranes. In *Proteinases in Mammalian Cells and Tissues* (Barrett, A. J., ed.) pp. 393–444. Elsevier, Amsterdam.

142. Weaver, L. H., Kester, W. R. & Matthews, B. W. (1977). A crystallographic study of the complex of phosphoramidon with thermolysin. A model for the presumed catalytic transition state and for the binding of extended substrates. *J. Mol. Biol.* **114**, 119–132.

143. Roques, B. P., Fournie-Zaluski, M. C., Soroca, E., Lecomte, J. M., Malfroy, B., Llorens, C. & Schwartz, J. C. (1980). The enkephalinase inhibitor thiorphan shows antinociceptive activity in mice. *Nature (London)* **228**, 286–288.

144. Roques, B. P., Lucas-Soroca, E., Chaillet, P., Costentin, J. & Fournie-Zaluski, M. C. (1983). Complete differentiation between enkephalinase and angiotensin converting enzyme by retro-thiorphan. Proc. Natl. Acad. Sci. USA 80, 3178–3182.
145. Bouboutou, R., Waksman, G., Devin, J., Fournie-Zaluski, M. C. & Roques, B. P. (1984). Bidentate peptides: highly potent new inhibitors of enkephalin degrading enzymes. Life Sci. 35, 1023–1030.
146. Umezawa, H. & Aoyagi, T. (1977). Activities of proteinase inhibitors of microbial origin. In Proteinases in Mammalian Cells and Tissues (Barrett, A. J., ed.) pp. 636–662. Elsevier, Amsterdam.
147. Okuda, M. & Arakawa, K. (1985). Potent inhibition of angiotensin-converting enzyme by [des-Pro³]-bradykinin or "converstatin" in comparison with captopril. J. Biochem. (Tokyo) 98, 621–628.
148. Hazum, E., Chang, K.-J. & Cuatrecasas, P. (1979). Specific nonopiate receptors for β-endorphin. Science 205, 1033–1035.
149. Shavit, Y., Lewis, J. W., Terman, G. W., Gale, R. P. & Liebeskind, J. C. (1983). Opioid peptides mediate the suppressive effect of stress on natural killer cell cytotoxicity. Science 223, 188–190.
150. Nilsson, J., von Euler, A. M. & Dalsgaard, C. J. (1985). Stimulation of connective tissue cell growth by substance P and substance K. Nature (London) 315, 61–63.
151. Rozengurnt, E., Legg, A. & Pettican, P. (1979). Vasopressin stimulation of mouse 3T3 cell growth. Proc. Natl. Acad. Sci. USA 76, 1284–1287.
152. de Souza, E. B., Whitehouse, P. J., Kuhar, M. J., Price, D. L. & Vale, W. W. (1986). Reciprocal changes in corticotropin releasing factor (CRF)-like immunoreactivity and CRF receptors in cerebral cortex of Alzheimer's disease. Nature (London) 319, 593–595.

Fluorescence Analysis of Protein Dynamics

A. P. DEMCHENKO

A.V. Palladin Institute of Biochemistry, Academy of Sciences of the Ukraine, Kiev 252030, USSR

I. Introduction 120
II. Intramolecular Motions in Proteins 122
    A. Distribution of Dynamic Microstates 122
    B. Analysis of Motions with Time Resolution 125
III. Dynamic Information Based on Fluorescence Data . . . 126
    A. Fluorescence Quenching 126
    B. Rotation of Aromatic Groups 128
    C. Dynamic Reorientation of Dipoles in the Chromophore
       Environment 129
IV. Fluorescence Spectroscopy of Molecular Relaxations:
    Theory 131
    A. Bakhshiev–Mazurenko model 131
    B. Model Based on Inhomogeneous Broadening of Spectra 132
V. Experimental Studies of Proteins by Fluorescence Spectro-
    scopy of Molecular Relaxations 137
    A. Relaxational Shift of Spectra 137
    B. Time-resolved Spectroscopy 137
    C. Red-edge Excitation Spectroscopy 138
    D. General Approach 142
VI. Intramolecular Dynamics of Dipolar Groups and Protein
    Function 143
    A. Charges and Dipoles in Protein Molecules 144
    B. Functional Consequences of Microstate Distributions . 146
    C. Molecular Relaxations in Enzyme Catalysis, Molecular
       Recognition and Allostery 147
VII. Outlook 150
    References 151

I. Introduction

Ultraviolet fluorescence has found very wide application in studies of proteins. The aim of this review is to analyse its possibilities and achievements in investigations of intramolecular dynamics in proteins at equilibrium and to discuss the significance of the results obtained and emerging in the comprehension of the mechanics of biochemical reactions in which proteins

ESSAYS IN BIOCHEMISTRY Vol. 22
ISBN 0 12 158122 5

participate. Fluorescence emission occurs spontaneously within nanoseconds of excitation, allowing us to achieve directly high time resolution (nanosecond and subnanosecond) in studies of the rates of different phenomena which are associated with changes of spectroscopic parameters in times comparable with those of the emission. This may be achieved both by direct recording of time-resolved spectra and by variations in time-averaged fluorescence parameters (spectral shifts, polarization or intensity) as a function of the different factors affecting the emission rate. Since fluorescence decay in the simplest case of excitation of a homogeneous ground-state population of chromophores which exhibit no excited-state reactivity is a mono-exponential function of time t:

$$I(t) = I_0 \exp\left(-t/\tau_F\right) \qquad (1)$$

then the excited state lifetime, τ, is widely employed to describe the emission rate. Time resolved studies of associated phenomena are thus easy if their characteristic times τ are of the order of the excited state lifetime, $\tau \approx \tau_F$. If $\tau \ll \tau_F$ such studies are more difficult but may be carried out in some cases by picosecond spectroscopy. In the case where $\tau \gg \tau_F$, no direct kinetics studies are possible, and only some limiting estimates can be made. Thus, the fluorescence method has considerable merit in the study of molecular dynamics. Its weaknesses result in general from the low structural resolution available. This last may be at least partially overcome if the spectroscopic properties of the chromophore and the reactions of its electronically excited states with the environment are well characterized through data obtained in simple model systems.

The indole chromophore of tryptophan is the most important chromophore in proteins. Intensity maxima of the fluorescence spectra of tryptophan in proteins range from 308 nm for azurin to 350–353 nm for peptides without regular structure and denatured proteins. The important feature of indole and tryptophan in this respect is the high sensitivity of their fluorescence spectra to interactions with the environment, through which substantial spectral shifts may result. Both non-specific general solvent effects such as dipole–dipole interactions[1-3] and specific excited-state complexes (exciplexes)[3,4] are probably involved in these solvent-dependent shifts. In proteins, these include interactions with surrounding groups of atoms in the protein matrix, as well as with molecules of water of hydration. The excited-state lifetime of tryptophan in proteins usually ranges between 3 and 5 ns, but may be substantially reduced on interaction with quencher groups.

In addition to studies of intrinsic fluorescence, we note the implications of extending these to various fluorescence probes and labels. The merit of fluorescence probes is their versatility with respect to spectroscopic properties and to the structure of the protein sites to which they are attached or adsorbed. Thus with pyrene probes, for instance, the τ_F range is extended to a hundred or so nanoseconds, while with eosin probes both fluorescence and phosphorescence may be observed simultaneously. The labels may be covalently attached to different protein groups, and many fluorescent analogues of different enzyme substrates and cofactors have been synthesized. Among the most popular are aminonaphthalene sulphonates, a group of fluorescence probes with substantial medium-sensitive fluorescence shifts. These, like the intrinsic chromophore tryptophan, are applied in studies of protein dynamics but with the potential advantages of introducing the probe to the desired protein site, a longer τ_F and of observing spectra in the visible. Potential disadvantages, on the other hand, include non-specific or multiple binding and induced modifications of protein properties. The following discussion will mainly be concerned with studies of intrinsic fluorescence, but will include reference to the most important studies employing fluorescence probes for the analysis of protein dynamics.

There are three approaches to the intrinsic fluorescence analysis of intramolecular dynamics in proteins which are being actively developed at present (Fig. 1):

(1) The quenching of fluorescence of tryptophan residues buried inside the globular protein interior by quencher molecules diffusing within the protein matrix.[5-7]

(2) The analysis of rotational motions of the chromophores by monitoring fluorescence depolarization with nanosecond time-resolution[8] or by variation of τ_F.[2,5,9]

(3) The investigation of the reorientational dynamics of dipolar groups surrounding the chromophore following changes in chromophore dipole moments which occur on electronic excitation.

In the following discussion, the last approach will be considered in some detail, particularly with respect to recently developed methods of time-resolved and red-edge-excitation spectroscopy[1,2,10,11] which allow the dipolar reorientational relaxation time τ_R, an important molecular dynamics parameter, to be determined.

II. Intramolecular Motions in Proteins

A. DISTRIBUTION OF DYNAMIC MICROSTATES

Individual protein molecules of the same primary structure are usually considered to be indistinguishable from each other. The strictly sequential biosynthesis of proteins, the thermodynamics of their folding, and X-ray

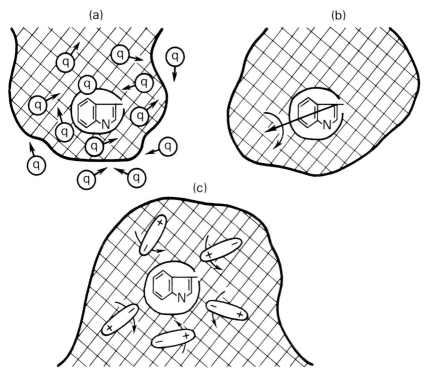

Fig. 1. Scheme illustrating the physical phenomena which are studied to obtain information on intramolecular mobility in proteins from fluorescence data. (a) Fluorescence quenching. The dynamic quenching of a fluorophore buried inside the protein globule required quencher diffusion inside the globule through densely packed groups of atoms. (b) Fluorescence polarization. Fluorophore rotation results in a decrease in the degree of polarization of emitted light. If the fluorophore is situated inside the globule, its rotation requires the motions of neighbouring groups. (c) Spectroscopic shifts and their red-edge effects. The mobility of dipolar groups surrounding the fluorophore results in a decrease of energy of emitted quanta and disappearance of the red-edge effects.

diffraction analysis of their crystals, all demonstrate that the protein molecule can be arranged into a single or at most a small number of native conformations. However, intrinsic motions within protein molecules must in general result in a certain divergence of properties between the nominally identical molecules in an ensemble. This means that, if time resolution is not considered, intramolecular dynamics reveals itself in non-identity of conformational states of different atoms and their groups and their variations within certain ranges, between individual molecules. Within the uniquely arranged protein conformational macrostate, there is a distribution of microstates which may be defined as a distribution in relative positions and orientations of the main chain and side groups within the same mean conformation. Such micro-disorder increases the configurational entropy contribution to the free energy of self-assembly, and favours the stabilization of the native macro-conformation.

In all experiments based on absorption or scattering of light or X-rays, the characteristic times of interaction of irradiation with matter are many orders of magnitude shorter than those of atomic motions. Therefore it is not the motion *per se*, but the disorder associated with it, that is being studied. Static disorder with no direct relation to mobility may be excluded by considering only the temperature-dependent component (Debye–Waller factor). The application of this approach to the results of high-resolution X-ray diffraction analysis of protein[12,13] have led to the demonstration of not only intramolecular mobility, but also variation of its level within a single protein molecule.

In studies of myoglobin, the protein structure in the haeme-binding site has been found to exhibit only small mean-square displacements of its atoms and can therefore be considered as an "aperiodic solid", while in regions distant from the binding site, mean displacements of 0.4–0.25 Å are found and the structure is thus closer to that of a liquid. The distribution of the haeme iron atom displacements in myoglobin has been studied in detail by Mössbauer spectroscopy,[14] and analysis of the temperature dependence revealed several distinctly different dynamic processes.[13] The distribution of microstates explains the thermal effects in NMR ring-current shifts,[15,16] and the dependence on temperature of the asymmetry of tryptophan environments in proteins which is observed in circular dichroism spectra.[17] The fine shifts of absorption spectra of intrinsic tryptophan residues (to longer wavelengths with temperature rise) may be observed for native proteins in solution by difference spectroscopy (thermal perturbation difference spectra).[1,18,19] These spectra demonstrate the occurrence in protein molecules of mobile (liquid-like) regions in which, on a rise in temperature, the distribution of microstates is shifted towards those having greater energies of interaction of the tryptophan ring with surrounding groups.

Optical spectroscopy is a unique method for observing the distribution of microstates differing in the energy of their interaction with the environment in the ground and excited states, using selective laser excitation of phonon-free bands in the electronic spectra characteristic of these microstates (site-selective spectroscopy). Some studies of this kind have already been made.[20,21] However, the observation of phonon-free lines requires liquid helium temperatures. At higher temperatures, in addition to the component responsible for inhomogeneity of the chromophore–environment interaction (inhomogeneous broadening), there arises a strongly temperature dependent component due to the interactions of electrons with vibrations and phonons (homogeneous broadening). For native proteins in solution, the broadening of absorption spectra may be observed easily by derivative spectroscopy,[1] but there is no possibility of differentiating these two components. However, it turns out that inhomogeneous broadening displays itself in a number of spectroscopic phenomena which are known collectively as "red-edge effects". These may be considered as kinds of site-selection spectroscopy in which broad electronic bands are photoselected instead of narrow phonon-free lines. The simplest of the observed effects is the red-edge (long wavelength) excitation-induced shift of fluorescence spectra to longer wavelength.[22,23] As shown in Section IV.C, this effect allows us not only to demonstrate microstate distributions in proteins, but also to observe the kinetics of their rearrangement.

B. ANALYSIS OF MOTIONS WITH TIME RESOLUTION

Various methods may be used to study the characteristic time evolution of the motions of small molecules in liquids. These usually occupy the picosecond time-range, and only those which require high activation energies occur on the nanosecond time-scale. By contrast the motions in protein molecules may be retarded by many orders of magnitude. This is because the kinetic units are segments of different sizes situated in the environment of other segments with high packing density. The motions are thus associated with overcoming high energetic and entropic barriers.

There exists no single method which covers the whole range of interest— from picoseconds to seconds. In addition, a proper level of structural resolution is required in these studies. It is not enough to say that there are motions with certain frequencies; it is important to know with which groups these motions are associated and what their mechanism might be.

In this connection results obtained by the NMR method, which allows observation in time of characteristic motions with high structural resolution, are very important. In proton magnetic resonance, changes in resonant frequency of aromatic protons due to exchange between different environments are usually observed. However, this allows only certain types of motions to be analysed, for example the ability of aromatic rings to undergo 180° flips can be studied in small proteins, such as pancreatic trypsin inhibitor, with a time resolution of 10^{-2}–10^{-4} s.[24] In ^{13}C-NMR, transverse and longitudinal relaxation times and nuclear Overhauser enhancement, which depend on nuclear motions, are observed with a time-resolution reaching 10^{-12} s. With this method, the motions of aliphatic side groups on a time-scale of 10^{-11}–10^{-12} s are indicated.[25] The recently developed two-dimensional NMR methodology provides a powerful tool for studying local geometry and flexibility of proteins,[26] but it is limited to peptides and proteins of small size. Paramagnetic probes are also widely employed in EPR studies of molecular dynamics. It is worth noting that, in the NMR and EPR methods conventionally employed in protein research, the temporal resolution is not usually direct, i.e. kinetics are not recorded in real time although time resolution down to 10^{-6} s is now available for EPR. Light emission spectroscopy (fluorescence and phosphorescence) is the method which has traditionally provided such a possibility. Here the absorption of a quantum of light occurs in times much shorter than any molecular motions, and the process of emission is delayed by many orders of magnitude to times which correspond closely to the time-range of the kinetic processes of interest. For phosphorescence, the microsecond to second time-range is available and phosphorescence quenching and spectral shifts associated with dynamics in proteins are being investigated with increasing interest.[28] The time-window of the fluorescence method is nanoseconds, that is from 10^{-10} to 10^{-7} s.

Although a nanosecond is an extremely short time in terms of normal human experience, on a molecular level it is not so short and is a very interesting time-span. A nanosecond is a very long time relative to the time-scales of intramolecular vibrations and rotational motions of small molecules in fluids. On the other hand, it is sufficient for translational diffusion of small molecules in fluid media over distances of several nanometres, i.e. of the order of several molecular dimensions. Rotational and vibrational motions may be retarded substantially and fall into the nanosecond time-range if they occur in viscous media or at the surface or in the interior of a protein matrix, and here molecular fluorescence relaxation and quenching are appropriate methods for their study. Rotational diffusion of whole protein molecules in aqueous solutions occurs in times of tens or hundreds of nanoseconds. This motion, as well as the more rapid intra-molecular rotation mobility of chromophore groups, may be studied by fluorescence depolarization. The elementary stages of biochemical reactions which are determined kinetically, however, occur on the microsecond to millisecond time-scale[29,30] and an intriguing question arises as to what extent the more rapid nanosecond mobility is required for protein functioning, and to what degree this may be involved in the elementary physical events that determine certain reaction stages.

III. Dynamic Information Based on Fluorescence Data

A. FLUORESCENCE QUENCHING

Methods of fluorescence quenching suggest an important approach to the analysis of intramolecular dynamics in proteins. Lakowicz & Weber[5] were the first to show that molecular oxygen is capable of quenching the fluorescence of even buried tryptophan residues and with unexpectedly high rate constants of $2–7 \times 10^9$ M^{-1} s^{-1}. They suggested that this quenching results from conformational fluctuations which allow oxygen molecules to diffuse within the protein matrix. Acrylamide is also capable of quenching the fluorescence of buried tryptophan residues, as has been clearly demonstrated for aldolase and ribonuclease T_1.[31] Furthermore, internal protein residues even exhibit noticeable quenching by ionic quenchers.[32]

Possible mechanisms for the effects of quenchers are usually discussed on the basis of two different models which are similar to those used in the treatment of kinetic hydrogen-exchange data.[33] One of these invokes relatively free diffusion of quenchers inside a viscous protein matrix. In this case, the quenching effect should have a small activation energy, and depend only slightly on solvent viscosity, but strongly on the size and charge of the quencher. The other model involves large-scale fluctuations and local unfolding of the protein. This mechanism of quenching should result in a high activation energy, and strong dependence on solvent viscosity, but only a small dependence on the size and charge of the quencher. In the case of oxygen, the observed activation energy is small ($3–4$ kcal mol^{-1})[5] and there is no significant dependence of the quenching rate on solvent viscosity.[34] The small size of oxygen molecules, absence of charge and good solubility in non-polar media suggest that they penetrate directly into the protein molecule without any obligatory protein-opening. For other quenchers, the most probable mechanism does involve local protein-opening. In the acrylamide quenching of aldolase and ribonuclease T_1 the activation energy is rather high ($9–11$ kcal mol^{-1}).[31] The efficiency of quenching by ionic quenchers and acrylamide is a function of T/η, the ratio of temperature to viscosity for the solvent.[32] The buried tryptophan residues of azurin,[31] alcohol dehydrogenase and alkaline phosphatase[35] are practically unquenched by acrylamide. Thus, easy penetration of small molecules (except of oxygen) and their subsequent diffusional motion within the protein molecule is barely possible; this requires extended fluctuations within the protein globular matrix.

The protein molecules themselves contain groups of atoms capable of quenching the fluorescence, and it is this fact that principally determines the considerable variation observed in protein fluorescence quantum yields and lifetimes. The most effective quenchers are carbonyls of the peptide chain, disulphide groups and amino, carboxylic and imidazole groups of the side-chains. The quenching may be both static and dynamic, the latter requiring nanosecond mobility of the chromophore environment to allow the dynamic collision of chromophore and quenching groups. The dynamic component of quenching may be revealed by observing the effects of temperature on protein fluorescence quantum yield, Q, and lifetime, τ_F.

Studies by Bushueva *et al.*[36] showed that, for an extensive number of proteins, the dependence of $1/Q$ and $1/\tau_F$ on temperature is linear in T/η, irrespective of whether the chromophore is at the surface or buried. The quenching is probably governed by some general feature or protein structure, whose dynamics are associated with solvent mobility.[37,38] The relationship of temperature-quenching parameters to the dynamic properties of the chromophore environment is not clear. Due to high local viscosity and short excited-state lifetimes, transient diffusion processes[39] may play an important role in these mechanisms.

Thus the data on dynamic fluorescence quenching in proteins provide direct proof for the existence of nanosecond intramolecular motions. However, the exact relationship between the parameters of quenching and the parameters describing protein dynamics cannot currently be extracted from these experiments.

B. ROTATION OF AROMATIC GROUPS

Information on rotational motions of protein molecules, as well as of chromophore groups within them, can be obtained by fluorescence polarization studies. On excitation by polarized light the chromophores with particular orientations relative to the electric vector of the incident light are preferentially photoselected, and the subsequent emission will tend to be less polarized the greater the rotational mobility of the excited chromophores. Three versions of the method are commonly employed:

(1) Recording polarization changes in real time,[2,8]

(2) Measuring steady-state polarization as a function of quenching by oxygen[40] or acrylamide[41] (shortening of τ_F);

(3) Studying the dependence of polarization on T/η, applying the Perrin equation.[42]

There is, where relevant, some difficulty in differentiating the effects of rotations of the protein as a whole and of the chromophores themselves relative to their protein environment. The intramolecular motions are thought to be more rapid and to be revealed as short-lived components in anisotropy decay curves, or as deviations from linearity of Perrin plots at high viscosities or low temperatures.[42]

The results hitherto obtained indicate that intramolecular rotations of tryptophan residues on the nanosecond time-scale may occur, but are not frequently observed, and their mechanism may not always be that of free diffusion in a viscous medium. Among native proteins with an internal tryptophan location it may only be azurin that displays large amplitude tryptophan rotations.[8] The results of studies on several other proteins agree with the model of nanosecond rotational relaxation with small angular amplitude (up to $-30°$).[41,42] In the cases where such motions occur, measurements of steady-state polarization in viscous solvents allow the determination of the thermal coefficient of frictional resistance to rotation.[43] This index displays temperature-dependent breaks, indicative of the activation of new rotational degrees of freedom.

A characteristic of any method based on fluorescence analysis is that the chromophore in its electronic excited state is not in equilibrium with its environment, and the structural relaxation of chromophore–environment interactions may be expected to influence the rotational relaxation of the chromophore. The nature of this possible influence is currently under discussion. Lakowicz[44] has suggested that the chromophore rotational motions are retarded in the course of relaxation. The possibility that induced chromophore rotational motions are associated directly with the relaxation has also been considered.[1] A possible way to differentiate normal Brownian and induced chromophore rotation is to determine whether the fluorescence polarization is affected by changes in excitation wavelength. While tryptophan emission has been excited over a range of wavelengths as long as 310 nm,[45] there are as yet no systematic studies on the effect of excitation wavelength.

C. DYNAMIC REORIENTATION OF DIPOLES IN THE CHROMOPHORE ENVIRONMENT

Dipole–dipole interactions of an excited chromophore with surrounding protein groups or solvent molecules produce substantial effects on the position of its fluorescence spectrum. Since the interaction energy depends on the magnitude and relative orientation of the dipoles, and on excitation the chromophore dipole moment changes both its magnitude and direction, the fluorescence spectra are directly related to chromophore environmental dynamics. Their position depends on whether the surrounding dipoles can be reoriented to a new equilibrium with the excited-state dipole. The emission is shifted to longer wavelengths during the course of relaxation.

According to the theory of relaxational shifts,[47,48] the magnitude of this effect (which is expressed as the change in emission frequency $\Delta\nu$) is proportional to the square of the vectorial difference between the chromophore dipole moments in the ground and excited states ($\mu_e - \mu_g$)^2, is a function of the dielectric constant ε_0 and refractive index, n, of the medium, and depends on the relaxation time, τ_R:

$$\Delta\nu \sim (\mu_e - \mu_g)^2 \times \left(\frac{\varepsilon_0 - 1}{\varepsilon_0 + 2} - \frac{n^2 - 1}{n^2 + 2}\right) \times [1 - \exp(-t/\tau_R)] \quad (2)$$

Here, the term which depends on the dielectric constant determines the spectral shift due to dipole–dipole interactions. This effect will be smaller, the greater the electronic polarization of the medium, which is expressed by the term in refractive index, n.

For fluorescence of indole and tryptophan in model polar media, the relaxational shift is of the order of 25–30 nm.[1] This is the major factor which, along with environmental polarity and ability to form specific complexes in the excited states (exciplexes), governs the position of fluorescence spectra in proteins. Some fluorescent probes display even greater relaxational shifts.

The relaxation times of dipolar molecules or groups of atoms in condensed media may be determined by the dielectric dispersion method, based on reorientation of dipoles in the external electric field.[46] Being without structural resolution, this method is practically useless in studies of intramolecular dynamics in proteins. On the other hand, using the fluorescence method, one may obtain selective information on mobility in the environment of tryptophan residues or in probe-binding sites. As a result of lack of information on the detailed description of relaxation process, both these methods rely on the Debye approximation. The real polar molecule is approximated to a sphere of radius a, which rotates in a continuous medium of viscosity η. The dipolar relaxation time is then expressed by

$$\tau_{\mathrm{d}} = 4\pi\eta a^{3}/kT \tag{3}$$

As distinct from dielectric relaxation, the dipolar relaxation manifested by fluorescence spectral relaxation does not directly reflect the rotational relaxation of the dipolar molecules themselves, but depends on the relaxation of the reactive field created by them acting on the excited chromophore. The relaxation time τ_{R} of the latter process may be expressed in terms of τ_{d} by the following approximate relation:[47]

$$\tau_{\mathrm{R}} \approx \frac{n^{2} - 2}{\varepsilon_{0} + 1} \tau_{\mathrm{d}} \tag{4}$$

For strongly polar liquids

$$\tau_{\mathrm{R}} \approx 0{\cdot}1\tau_{\mathrm{d}}$$

The dipolar-reorientational relaxation time is an important parameter describing intermolecular and intramolecular dynamics. In the following sections we shall discuss ways of obtaining this parameter from data on spectroscopic shifts and analyse the results of protein studies.

IV. Fluorescence Spectroscopy of Molecular Relaxations: Theory

The absorption of a quantum of light produces instantaneous perturbation of the equilibrium between the chromophore and its environment. Relaxation then occurs towards the establishment of a new orientation at a distribution of dipoles which is in equilibrium with the excited state. This distribution may be attained or not attained, depending on the relative values of the fluorescence lifetime, τ_F, and the relaxation time, τ_R.

A. BAKHSHIEV–MAZURENKO MODEL

Consider the four-level energy diagram (Fig. 2) which usually serves as a basis for discussion of the effects of intermolecular interactions and relaxations in electronic spectra. The absorption of a quantum of light, $h\nu_a$, by the system results in its transition to the excited Franck–Condon level, E_e^{FC}, which is not at equilibrium with the environment. Quanta emitted during the initial very short period of time ($t \simeq 0$) are emitted from this level, their energy ($h\nu_0^F$) is high, and the fluorescence spectrum is located in a low

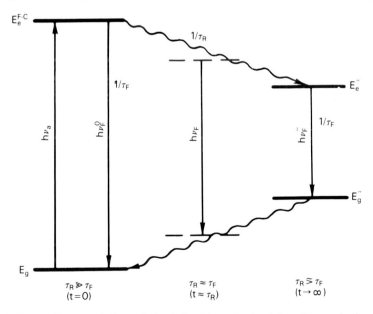

Fig. 2. Energy diagram of electronic levels for intermolecular interactions and relaxations (continuous relaxation model).

wavelength range. This case may be observed in the steady-state spectra of solid solutions, for which $\tau_R \gg \tau_F$. If the observation time is very long ($t \rightarrow \infty$) or $\tau_R \ll \tau_F$, which is the case for a liquid solvent of low viscosity, the rearrangement of the surrounding dipoles will have been complete, and the excited chromophore will have reached equilibrium with the environment prior to emission.

The Bakhshiev–Mazurenko model[47–49] considers the case when emission decay and solvent relaxation are commensurate, that is, a decrease in the number of excited-state chromophores occurs at a similar rate to orientational relaxation of the environment. In the Debye approximation (eqn 3) the mean frequency of emission $\bar{\nu}$ depends on τ_R and τ_F in the following way:

$$\frac{\bar{\nu} - \bar{\nu}_\infty}{\bar{\nu}_0 - \bar{\nu}_\infty} = \frac{\tau_R}{\tau_R + \tau_F} \qquad (5)$$

where $\bar{\nu}_0$ is the mean emission frequency of the unrelaxed chromophore, which is observed in viscous media at early times or at low temperatures, and $\bar{\nu}_\infty$ is the frequency of emission of the fully relaxed excited chromophore which occurs at high temperatures or at long times, in such solutions.

Equation (5) represents quite a good approximation for the description of relaxational shifts of spectra due to variations in τ_R and τ_F brought about by changes in temperature and by fluorescence quenchers. It allows the values of τ_R to be estimated, provided τ_F, $\bar{\nu}_0$ and $\bar{\nu}_\infty$ are known for the chromophore–environment system under study.

B. MODEL BASED ON INHOMOGENEOUS BROADENING OF SPECTRA

Systems with slow relaxation, i.e. those for which $\tau_R > \tau_F$ display specified "edge" effects in excitation, in emission and in transfer of excitation energy.[1,22,23,50] The analysis of these effects requires more complex models accounting for the statistical distribution of the interaction energy of molecules with the environment, and describing selection by the exciting or emitted light quantum of species within the chromophore population whose excitation or emission energy respectively diverges from the mean. The chromophore–environment interactions may differ in their mutual orientation of dipoles and their interaction energy, resulting in differences in the electronic transition frequencies. The spectroscopic behaviour of the system is determined by this distribution at the time of excitation and its subsequent changes in time (the relaxation process).

Consider the energy level diagram (Fig. 3) which accounts for the distribution of interaction energies.[1] The energy E of any electronic level is a sum; $E = E_0 + \Omega(W_{dd})$, where E_0 is the energy level in the absence of

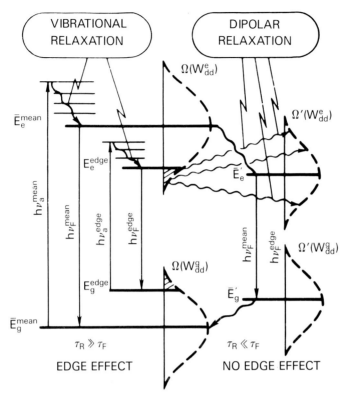

Fig. 3. Energy diagram of electronic levels accounting for the distribution of interaction energy in the ground and excited states and intermolecular relaxations (inhomogeneous broadening, continuous relaxation model).

interaction, and the stabilization energy is represented by a distribution $\Omega(W_{dd})$. If excitation is produced by low energy quanta $h\nu_a^{edge}$ (occurring at the long-wavelength edge of the absorption spectrum), there occurs a photoselection of only those chromophores whose interaction energy with the environment is minimal in the ground state and maximal in the excited state. The photoselection is based on the fact that a quantum of light with low energy can be absorbed only by those chromophores for which $h\nu_a^{edge} = E_e^{edge} - E_g^{edge}$. For the rest of the chromophores, having a wider gap between the ground and excited states, the energy of the quantum is insufficient for it to be absorbed.

If during the lifetime τ_F essentially no dipolar-reorientational relaxation has taken place, the energy of the emitted quanta is shifted to longer wavelengths compared with those of quanta emitted under more usual excitation conditions, and the red-edge effect occurs. If $\tau_R \ll \tau_F$, the dipolar relaxation results in rapid redistribution of interaction energy with the environment. As a result, the chromophore which is excited by a low-energy quantum at the red-edge "forgets" this, and the emission spectrum becomes independent of the excitation wavelength. Thus, information on structurally non-equilibrium excited states may be obtained by observing the excitation wavelength dependence of steady-state spectre.

For the description of red-edge effects as a function of relaxational properties of the medium we use the Bakhshiev–Mazurenko model and consider relaxation as a process of establishing equilibrium in an ensemble of interacting species. Then eqn (5) is valid for both mean and edge excitations. Taking into account the fact that $\bar{\nu}$ values will depend on excitation wavelength, and $\bar{\nu}_\infty$ values will not, we obtain:

$$\frac{\bar{\nu}^{\text{mean}} - \bar{\nu}^{\text{edge}}}{\bar{\nu}_0^{\text{mean}} - \bar{\nu}_0^{\text{edge}}} = \frac{\tau_R}{\tau_R + \tau_F} \qquad (6)$$

Equation (6) may be applied easily to the estimation of τ_R. As distinct from eqn (5), it does not contain $\bar{\nu}_\infty$, the position of the spectrum of the completely relaxed state. The determination of the latter is a substantial problem in experiments with proteins, since τ_F is small and application of high temperatures is limited by the narrow range of stability of the native protein structures.

Thus, in the method of edge-excitation spectroscopy, along with the determination of τ_F it is sufficient to record fluorescence spectra at two excitation frequencies and to determine the shift between them, $\bar{\nu}^{\text{mean}} - \bar{\nu}^{\text{edge}}$, under the conditions of the experiment, and $\bar{\nu}_0^{\text{mean}} - \bar{\nu}_0^{\text{edge}}$ in the limiting case of early times or low temperatures. The condition $\tau_R = \tau_F$ is satisfied at the half-transition point, when

$$\bar{\nu}^{\text{mean}} - \bar{\nu}^{\text{edge}} = \tfrac{1}{2}(\bar{\nu}_0^{\text{mean}} - \bar{\nu}_0^{\text{edge}}) \qquad (7)$$

The magnitude of the effects of red-edge excitation on the fluorescence spectra of tryptophan in solutions of different viscosity is illustrated in Fig. 4. In Fig. 5 are presented the results of the determination of τ_R for a model viscous solvent (glycerol), with indole and tryptophan as spectro-scopic probes, conducted by two methods: the first based on the Bakhshiev–Mazurenko model (eqn 5) and the second, suggested by the author, based on eqn (6) using data on edge-excitation shifts. A satisfactory correlation of these results is observed with consistent differences in τ_R not exceeding a factor of 2–3 (which is in order with the assumptions made). The range of relaxation times determined covers two orders of magnitude. The spec-troscopic results show reasonable correlation with data obtained by the dielectric dispersion method.

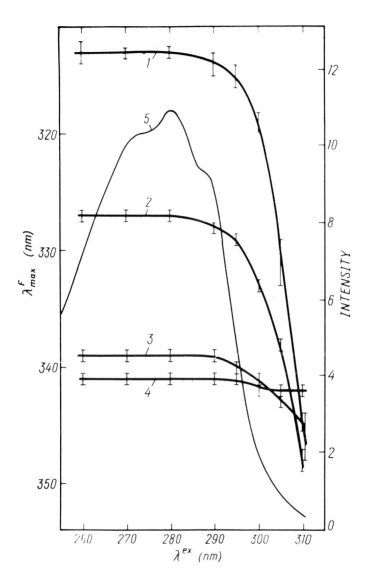

Fig. 4. Dependence of the wavelength of fluorescence maxima on excitation wavelength for tryptophan in glycerol at temperatures of: (1) −196°C; (2) −14°C; (3) 20°C; and (4) 50°C. (5) is the absorption spectrum of tryptophan in glycerol at 20°C.

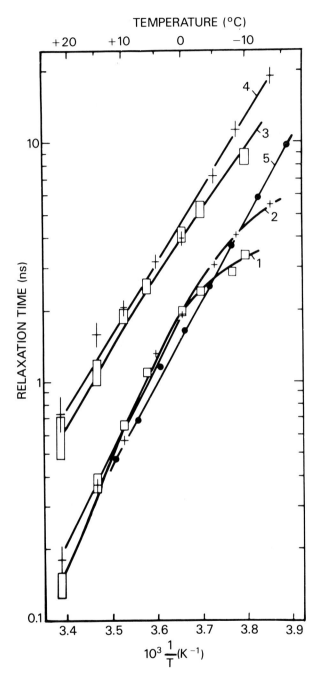

Fig. 5. Dependence of dipolar reorientational relaxation time on temperature for indole and tryptophan in glycerol. Data at $\lambda^{ex} = 270$ nm for indole (1) and $\lambda^{ex} = 280$ nm for tryptophan (2) calculated using eqn (5). Data at 270 and 295 nm for indole (3) and 280 and 305 nm for tryptophan (4) calculated using eqn (6). (5) is the temperature dependence of glycerol dipolar relaxation time obtained from dielectric studies using eqn (4).

V. Experimental Studies of Proteins by Fluorescence Spectroscopy of Molecular Relaxations

A. RELAXATIONAL SHIFT OF SPECTRA

It follows from eqn (5) that, by variation of τ_R and τ_F, the limiting short-wave and long-wave positions of fluorescence spectra may be obtained. These data, together with fluorescence lifetimes and positions of spectra determined under experimental conditions of interest, allow one to determine τ_R. Variation of τ_R and τ_F may be achieved in two ways: either by changes in temperature or by introduction of dynamic fluorescence quenchers.

There are substantial difficulties in the interpretation of temperature-dependent shifts of spectra due to the thermal lability of proteins. The low-temperature conditions, however, can be accessed either by addition of glass-forming co-solvents[27,52] or directly by freezing in water.[53] Relaxational shifts were observed in studies of riboflavin–chymotrypsinogen[52] and eosin serum albumin[27] systems, suggesting relaxational motions in protein-binding sites. Permyakov & Burstein[53] observed that the low-temperature shifts in several proteins differ significantly from that expressed by eqn (5), which probably results from co-operative immobilization of intrinsic mobility with the participation of the aqueous solvent. Thus, proteins with exposed tryptophan residues (cobre neurotoxins, serum albumin, papain) show spectral shifts of 5–12 nm in the temperature range from 0 to $-20°C$ and some further shifts in the -20 to $-90°C$ range. For proteins with buried tryptophan residues (trypsin, chymotrypsin, β-lactoglobulin), a 2–5 nm shift is revealed in the -20 to $-90°C$ range only.

Fluorescence quenching to vary τ_F has been suggested as a method for obtaining "lifetime-resolved fluorescence spectre".[2,6] A shift of spectra as a function of τ_F indicates dipolar relaxation. Earlier Lakowicz & Weber[5] observed short wavelength shifts of fluorescence spectra on quenching with oxygen. Since these observations were made on proteins with several tryptophan residues per molecule, the shifts might also have originated from selective quenching of long-wave-emitting residues. Several single-tryptophan proteins have been found to display spectral shifts on acrylamide quenching,[54] but no attempts were made in those studies to determine τ_R. If under initial conditions $\tau_R \approx \tau_F$, the quenching experiments do not allow $\bar{\nu}_\infty$ to be determined.

B. TIME-RESOLVED SPECTROSCOPY

If emission and relaxation are associated in time, then by studying time-resolved spectra we can observe, in principle, the spectrum of non-relaxed states at initial times, the spectra of partially relaxed states as time goes on, and the limiting spectrum whose position corresponds to the completely relaxed state.[48,55] If the decay curves are observed at different emission wavelengths, the decay is not exponential: additional fast positive component(s) of the decay exist(s) on the short wavelength edge, and fast negative component(s) in the long wavelength region.[47,48] The mean excited-state lifetime increased on observation from shorter to longer wavelengths.[2] The spectral and temporal inhomogeneity of emission is also revealed in phase fluorimetric studies: the phase angle and depth of modulation become a function of emission wavelength.[2,47,48]

The results show that while in liquid solvents tens of picoseconds are sufficient to equilibrate the dipoles,[56] relaxation in protein molecules is substantially retarded. Nanosecond dynamics of spectra were observed in apohaemoglobin and bovine serum albumin[57–59] labelled with 2-(p-toluidinylnaphthalene)-6-sulphonate (2,6-TNS). In studies of intrinsic protein fluorescence, a dependence of decay kinetics on emission wavelength typical of dipolar relaxation was observed for chick pepsinogen by Grinvald & Steinberg.[60] These authors succeeded in detecting a component with negative amplitude in the long wavelength of region tryptophan emission. However, in subsequent studies of emission decay in 17 other proteins, eight of which contained only a single tryptophan residue, these authors failed to find such a component.[61] These results do not exclude the possibility of nanosecond relaxation, which may be masked by other processes and require some special conditions for unequivocal observation. It is noteworthy that seven out of the eight single-tryptophan proteins (the exception being azurin) exhibit multi-exponential decay kinetics. It is quite likely that, at the moment of excitation, the sole tryptophan residue is found in different conformational microstates, each with its own probability of excited-state deactivation, and the observed excited-state lifetime is an average value. In the presence of short-lived positive components in the emission, the short-lived negative component at long wavelengths may be masked. An increase in the mean lifetime in the long wavelength region of emission[60,62] may be considered as a probable indication of nanosecond dipolar relaxation.

Thus, applying fluorescence spectroscopy with nanosecond time-resolution, one can observe motions within protein molecules occurring on the nanosecond time-scale. In several cases the results demonstrate that tryptophan residues or fluorescence probes are situated in a "micro-viscous" medium with substantial nanosecond mobility.

C. RED-EDGE EXCITATION SPECTROSCOPY

Red-edge excitation spectroscopy is a study of the dependence of fluorescence spectra on excitation wavelength at the long-wave excitation edge.

According to the theory (Section IV.B) such excitation results in a shift of fluorescence spectra to longer wavelengths in the case where the relaxational rate is slower than the excited-state decay.

Initial studies on the fluorescence probe 2,6-TNS associated with proteins have given unexpected results. This probe, as well as other aminonaphthalene sulphonates (ANS, DNS),[57] was considered earlier as an indicator of the polarity of binding sites. The observations by Demchenko[51] of substantial edge-excitation fluorescence shifts demonstrate that the short-wave shifts observed in emission spectre of such probes on binding to proteins results not (or not only) from hydrophobicity of their environment, but also from the absence of dipolar-orientational equilibrium in the nanosecond time-range. Thus, the fluorescence shifts observed on change of excitation wavelength from 360 to 400 nm are 14 nm for β-lactoglobulin, 7 nm for β-casein, 8 nm for bovine serum albumin and 13 nm for human serum albumin.

The principal difference between probe behaviour in proteins as opposed to liquid media in which it exhibits the same emission spectrum, is that in the latter the spectrum is not shifted on edge excitation (Fig. 7). Since the absorption spectra of the probes are not changed on binding, heterogeneity of the binding cannot explain this effect. In addition, as Fig. 7 demonstrates, the width of the fluorescence spectrum is somewhat smaller, not larger, on edge excitation. We would not observe this in the case of an impurity in the probe or with heterogeneity of its binding.

Recently substantial red-edge excitation shifts of fluorescence spectra were observed for the complexes of 2,6-TNS with melittin[10] and apomyoglobin.[63] In the latter case, there were almost no differences between the results obtained at 20 and 4°C, which is in line with the results on other proteins studied,[51] which also showed a lack of substantial temperature dependence of the effect.

In studies of intrinsic protein fluorescence it is of interest to investigate first proteins with a single tryptophan residue per molecule, and to compare the magnitude of the edge effect with the positions of fluorescence maxima.[64,65] It was found that in proteins with the shortest wavelength emission (azurin, whiting parvalbumin, ribonucleases T_1 and C_2), the edge effect is not observed.[65] This may reflect the hydrophobic environment of the tryptophan indole ring and the small magnitude of inhomogeneous broadening of the spectra. There is also no observable effect in the case of very long-wavelength emitting proteins with fluorescence maxima at 340–350 nm (melittin at low ionic strength, TI-Ajl protease inhibitor, basic myelin protein and β-casein), but the reason for this is quite different: the tryptophan residues are exposed to the rapidly relaxing aqueous solvent. The edge-excitation fluorescence shift is clearly observed in studies of proteins emitting at intermediate wavelengths (human serum albumin in N- and

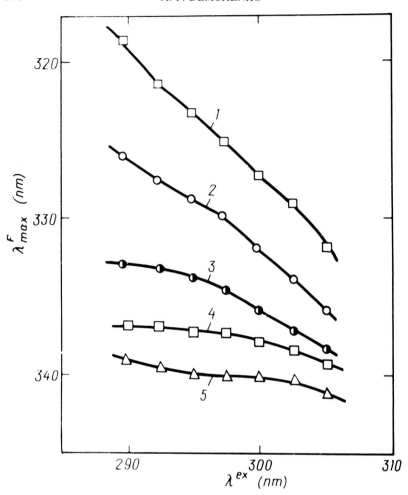

Fig. 6. Depe.idence of the position of fluorescence maxima on excitation wavelength for single-tryptophan proteins emitting in the medium wavelength range.[65] (1) Human serum albumin at pH 7·0 in the presence of 10^{-4} M sodium dodecylsulphate; (2) albumin at pH 3·2; (3) bee venom melittin at pH 7·5 with 0·15 M NaCl; (4) protease inhibitor from *Actinomyces junthinus* at pH 2·9; (5) albumin at pH 7·0.

F-forms, TI-Ajl protease inhibitor at pH 2·9, melittin at high ionic strength and serum albumin complex with sodium dodecylsulphate[64,65] (Fig. 6). This demonstrates the existence in these proteins of both a distribution of microstates with different orientations of dipoles in the tryptophan environment, and slow (on the nanosecond time-scale) mobility of these dipoles, resulting in emission from structurally in-equilibrium electronic excited state.

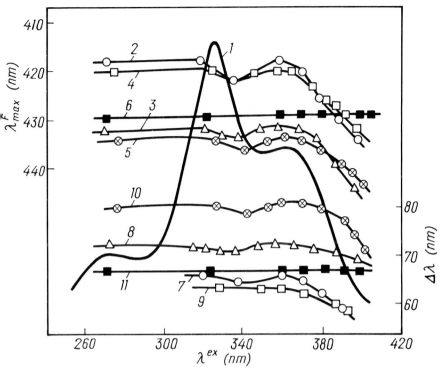

Fig. 7. Fluorescence excitation spectrum of 2-(p-toluidinylnaphthalene)-6-sulphonate (2,6-TNS) and the dependence of the maxima of its fluorescence specta λ^F_{max} (2–6) and bandwidths $\Delta\lambda$ (7–11) on excitation wavelength.[51] 2,6-TNS bound to β-lactoglobulin (2,7) and human serum albumin (3,8) in 0·05 M tris-HCl buffer, pH 7·0 at 20°C. 2,6-TNS in glucose glass at 20°C (4,9), in glycerol at +1°C (5,10) and in 80% aqueous ethanol at 20°C (6,11).

The property of serum albumin of binding different metabolites and some other substances, including detergents, is well known. The tryptophan residue is thought to be integral to the structure of the binding sites. Its fluorescence spectrum is substantially shifted to shorter wavelengths on binding such substances. In studies of the albumin complex with sodium dodecylsulphate, the fluorescence shift on changing to edge excitation at 305 nm reaches 11 nm,[64] which is greater than for any other single-tryptophan protein studied so far.[65] Under the conditions of our experiment the protein molecule is native and only the sites with high affinity for detergent are occupied. The substantial red-edge effect demonstrates that, in the complex, the tryptophan residue becomes immobilized, with slow relaxation of surrounding dipoles. A similar effect was observed for albumin associated with a fluorescence probe (Fig. 7). In neither case was a substantial effect of temperature in the range 0–40°C observed. It should be mentioned that the dipole relaxation properties of the tryptophan environment in a native protein are not homogeneous, and there may be differences of many orders of magnitude in the characteristic times of motions of specific groups. In the ligand-free form of albumin, a small red-edge effect[64] and substantial relaxational shift of spectra at sub-zero temperatures[27] are observed. This indicates high mobility of the dipolar groups in the absence of the ligand and their immobilization on ligand binding.

Among multi-tryptophan proteins, the most significant red-edge effect (quantitated as the difference between the maxima of fluorescence spectra on excitation at 290 and 305 nm) was observed for papain in active (13 nm) and inactive (10 nm) forms. Substantial shifts are also reported for aspartyl- and valyl-tRNA-synthetases from rabbit muscle (8 nm). For rabbit aldolase A the shift is 6 nm, skeletal muscle myosin 4·5 nm, chymotrypsin 2·5 nm and carbonic anhydrase 2 nm. For trypsin and β-lactoglobulin no shifts are detected.[1] For "normal" (290 nm) excitation, all the proteins mentioned above possess fluorescence spectra with a maximum at about 330 nm. Edge excitation adds a new dimension which allows delineation of differences in protein chromophores with respect to the dynamic properties of their environment.

D. GENERAL APPROACH

As seen from the above discussion the methods of red-edge excitation and time-resolved spectroscopy can both be utilized in the study of the same dynamic phenomenon, but the information obtained is not entirely equivalent. The methods, therefore, complement each other, and ambiguities in data interpretation may be overcome by their joint application. Cases in which relaxation times are much faster or much slower than the chromophore excited-state decay are difficult to observe by nanosecond time-resolved spectroscopy. Again, even if nanosecond relaxation of spectra is detected, it may not necessarily be associated with dipolar relaxation, but may reflect other excited-state complex (exciplex) formation. On the other hand, in red-edge excitation spectroscopy, cases in which $\tau_R \gg \tau_F$ are resolved easily, although structural heterogeneity in the ground state may lead to some difficulties. Red-edge excitation spectroscopy may of course be combined directly with time-resolved spectroscopy by recording time-dependent emission spectra or the emission-wavelength dependence of decay as a function of excitation wavelength.

Dipolar reorientational relaxation is probably the most rapid process which requires motions of atomic nuclei. The structural relaxation associated with translational motions in solution is much slower and occurs at times when the neighbouring dipolar groups are already at equilibrium.[56]

Thus, as distinct from dipolar relaxation, nanosecond kinetics of other slow processes should not depend on excitation wavelength, and this property may be used to resolve them.

In connection with the above discussion, we consider the results of two studies in which the fluorescence probe 2,6-TNS was utilized. One of them was aimed at investigation of probe mobility in a complex with melittin;[10] the other at mobility in the membrane phospholipid bilayer.[67] In the case of melittin no nanosecond kinetics of spectral shifts was observed. However, both steady-state spectra and decay curves depended strongly on the excitation wavelength, providing evidence for the rigidity of the probe-binding site on the nanosecond time-scale. The properties of the probe in the phospholipid membranes were quite different. The nanosecond relaxation of spectra is easily observed, but neither steady-state spectra nor the decay curves at different emission wavelengths were dependent on the excitation wavelength. This suggests that dipolar relaxation in this case is a rapid, sub-nanosecond process, and the slower, nanosecond dynamics is associated with some other phenomenon, probably translational motion of the probe out of the bilayer.

Thus, modern fluorescence molecular relaxation spectroscopy is able to report directly on molecular dynamics in the nanosecond time-range and to quantitate the results of reactions outside this time-range. The results obtained bear witness against simple-minded pictures of protein molecules as micro-liquids or micro-solids. There exists instead a spatial and temporal hierarchy of structural dynamic processes, some of which are so substantially slowed down that they occur in the nanosecond or even in the microsecond time-range.

VI. Intramolecular Dynamics of Dipolar Groups and Protein Function

The above results obtained by fluorescence methods, as well as numerous data derived using a variety of other techniques, bear witness that the concept of proteins as dynamic systems is sufficiently well based experimentally. There is no need for further verification of this concept unless it be aimed at the analysis of specific motional modes and their relevance to protein function. Molecular relaxation spectroscopy, in particular, is directed towards quantitating dipolar group motions in proteins. The principal conclusions which have already been reached[1,64,65,68] are the following:

(1) he spread of dipolar interaction energies of different microstates of the same site in a protein molecule may be as much as several kJ mol^{-1}.

(2) Relaxation of dipolar groups covers a very extended time interval and may occur or not occur in the nanosecond time-range.

It would appear to be of some interest to concentrate attention on charge and dipolar interactions and motions in protein molecules and to consider theoretical and experimental data on their role in protein functions.

A. CHARGES AND DIPOLES IN PROTEIN MOLECULES

Electrostatic interaction in proteins has not received much attention in textbooks, even though it is a very important factor in stabilizing native protein structure.[69-71] It differs from other factors (hydrophobic interactions and hydrogen bonding) in its long range of action. Electrostatic interactions are not limited to the charged group interactions which are often displayed by ionic pairs,[70] but include interactions of dipoles with ions and also dipole–dipole interactions.

The dipole moment of the peptide group is about 3·5–3·6 Debye units. On the formation of α-helices, the dipole moments of the peptide groups become oriented parallel to each other, resulting in the formation of macro-dipoles, which carry partial positive charge at the N-terminal and partial negative charge at the C-terminal ends of the chain.[72,73] As a result the α-helical segment becomes a macro-dipole with a dipole moment which may reach hundreds of Debye units. This is the reason why, in α-helical proteins, the energetically most favourable conformation is that having anti-parallel orientation of α-helical segments.[73] In anti-parallel β-structures, the peptide group dipole moments are perpendicular to the chain direction, and a favourable anti-parallel orientation of dipoles is set up by interactions of neighbouring groups which belong to different chains. This contributes substantially to the rigidity of this structure. In parallel β-structures, the dipoles are situated at an angle, and a certain uncompensated dipole moment exists.[73] Many amino acid residues possess dipolar properties, including those which are considered to be of low polarity, e.g. tyrosine and tryptophan with dipole moments of 1·55 and 3·4 Debye units, respectively.[1] Their position and orientation is determined to a considerable extent by compensation of the electric field of other charged and dipolar groups. Hydrophobic residues, on the other hand, are situated in clusters: their contact with polar groups is energetically unfavourable. This results in substantial variations in effective dielectric constant within a single protein globule, spanning up to two orders of magnitude.[74,75] Charges inside the globule are surrounded by dipolar groups.[76,71]

A scheme illustrating the interaction of charges and dipoles within a protein globule is presented in Fig. 8. As well as the protein dipoles, the participation of water in these interactions is essential. The external water of

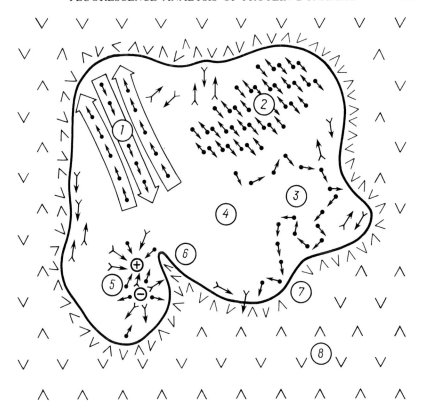

Fig. 8. Scheme illustrating the arrangement of protein and solvent dipoles. ●→ denotes the main chain dipoles; >→ denotes the dipoles of the side-groups; > denotes the water dipoles. (1) α-helical region: the main chain dipoles are oriented parallel to the chain direction; (2) region of anti-parallel β-structure: the dipoles are oriented perpendicular to the chain directions and the structure is stabilized by interchain dipolar interactions; (3) region of irregular arrangement of main chain and side-chain dipoles; (4) clusters of hydrophobic side chains with no permanent dipoles; (5) clusters of oriented dipoles surrounding the charged groups; (6) immobilized water molecules; (7) hydration water; (8) solvent water.

hydration, which amounts approximately to one monomolecular layer, differs from that of the bulk solvent in having a decreased mobility with correlation times of 10^{-10}–10^{-11} s.[77,78] In addition, there may be water molecules trapped within the globular interior,[69] and their mobility is governed by the surrounding protein structure.

Thus a hierarchy of dipoles exists in protein molecules which spans the whole range from groups of small size to macro-dipoles of α-helices. Their mobility may differ by several orders of magnitude. Dipole-reorientational equilibrium on electronic excitation may be established by both sterically limited rotations of small dipolar groups in the chromophore environment and by induced rotations of the chromophores themselves with respect to rigid dipoles. In proteins, the characteristic times for chromophore rotations (as determined from fluorescence polarization data) may have magnitudes similar to the relaxation times of the surrounding dipoles which are obtained from data on spectroscopic shifts. This situation is difficult to simulate in viscous chromophore solutions, where the rotation of solvent molecules is several orders of magnitude faster than chromophore rotations. The establishment of structural equilibrium in proteins may also require translation of charged and dipolar groups, and these motions may occur on the same time-scale as dipole-reorientational relaxations, whereas in model solvent media the latter process is several orders of magnitude faster.[79] Thus, fluorescence spectroscopic methods are capable of quantitating structural relaxation in proteins, in a broad sense, in terms of the establishment of a new structural equilibrium after an initial perturbation. According to the fluctuation–dissipation theorem,[80] the mean correlation time for spontaneous fluctuations in a system is directly related to the relaxation time observed after a small initial perturbation. Thus relaxation dynamics will reflect the true equilibrium dynamics within protein molecules.

Electrostatic interactions and their fluctuations have proved to be of considerable importance in the formation of specific intermolecular complexes, in molecular recognition, and in biocatalytic phenomena. The interacting molecules display complementarity of their electrostatic potentials.[69,70,81] Phosphate-binding sites in proteins are usually situated close to the N-terminal end of α-helical macro-dipoles.[73] These dipoles are thought to participate directly in the mechanism of papain protease activity[82,83] and in the hydrolytic action of phosphoglyceromutase.[84] Attempts have been made to probe the enzyme transition state by observing ionic strength effects and considering the dipole moment difference in the ground- and transition-state complex.[85] It is suggested that electrostatic substrate stabilization[86] and its screening from mobile solvent dipolar environments[87] are important factors promoting enzyme catalysis. Thus the presence of both dipolar and charged groups, and their motions in space and time, may determine the specificity of intermolecular interactions in proteins and protein systems.

B. FUNCTIONAL CONSEQUENCES OF MICROSTATE DISTRIBUTIONS

The results discussed above demonstrate that there exists a distribution of protein microstates differing in their interaction energies at the level of groups of atoms. It may be suggested that such distributions will influence the overall rate of some associated process if the rates of its elementary stages are faster than the transition rate between microstates. The microstates may be considered here as species with distinct intra- and intermolecular interaction properties. As a result of the distribution in activation energy[27] or activation entropy there will be a distribution in reaction rates, which can be detected as kinetic heterogeneity by reaction kinetics methods. If the rates of protein motions are fast compared to the reaction rates, kinetic homogeneity, averaged over all the microstates, will result.

Of importance in this respect are data on ligand rebinding haemoglobin after laser-induced dissociation.[88-90] At temperatures below 200–300 K non-exponential reaction kinetics is observed. Above this range, however, rapid exchange between conformational microstates results in exponential reassociation kinetics. The width of the apparent activation energy distribution is several kcal mol^{-1}. The steady-state pathway for oxygen or carbon monoxide binding by haemoglobin and myoglobin is shown to be absent, and the apparent activation energy is shown to be a function of intramolecular motions. A non-equilibrium distribution of microstates of ion-channel proteins in membranes may be important in the mechanism of transmembrane ion transport. It has recently been suggested by Läuger,[91] that such a distribution would result in macroscopic irreversibility of the ionic transport reaction.

In all probability, the distributions of microstates which are observed by reaction kinetics methods are of the same origin as those which result in inhomogeneous broadening of spectra and "red-edge" effects in molecular relaxation spectroscopy. This viewpoint should be capable of being proved directly in studies of photochemical reactions with proteins.

C. MOLECULAR RELAXATIONS IN ENZYME CATALYSIS, MOLECULAR RECOGNITION AND ALLOSTERY

The results on ligand association kinetics with haemoglobin[89,90] show a strong dependence of the rate constant not only on temperature, but also on solvent viscosity. The twisting photoisomerization rate of bilirubin bound to serum albumin in a high affinity site which is screened from solvent water also displays a strong solvent viscosity dependence.[92] In line with these results are the data obtained on Rayleigh scattering of Mössbauer emission,[14] which along with fluorescence quenching data demonstrate the strong dependence of intramolecular dynamics in proteins on solvent viscosity. Enzyme reaction rates may also depend on viscosity,[93] even when they are not limited by substrate or product diffusion rates. Solvent viscosity is thought to influence the rate of formation and dissociation of protein complexes through its influence on protein dynamics.[94] Thus the question arises, in what way can protein and solvent dynamics affect the rate of activated processes in proteins?

The problem of the influence of solvent molecule mobility on reaction kinetics is at the initial stage of its development, even in the chemistry of simple two-component reactions in liquid media. It is known, however, that non-equilibrium solvation,[95] solvent dipole translational diffusion[96] and orientational relaxation[97] can influence the reaction rates. The quantum mechanical theory of Dogonadze et al.[98] considers the solvent–solute interaction as one of the reaction coordinates, and the rate of medium reorganization is thought to influence the reaction rate directly. Still less clear are non-equilibrium and relaxational processes in enzyme catalysis. The accessibility of substrate or ligand-binding sites may fluctuate on the time-scale of the reaction[99,100] and the most important motions are probably those of rigid segments such as α-helices, which determine the size of the reaction clefts.[101] Transfer and dissipation of energy within protein molecules[102,103] and with solvent water[104] may occur through charge and dipolar interactions. The molecular dynamics simulations of a fluctuating enzyme–substrate complex carried out by Warshel[105] show that fluctuation in electrostatic potential is a key dynamical factor in proton transfer reactions. According to Dogonadze[106] protein dynamics is considered as an important degree of freedom associated with the enzyme reaction. The dynamics influences the reaction rates if the electronic states of reactants depend substantially on motion along the conformational coordinate. The difficulty of the problem is that the standard transition state theory formalism is unable to evaluate the experimental data in cases in which the dynamics of the reactant environment is involved. The data obtained by the Frauenfelder group[88,89] on oxygen binding by haemoglobin and myoglobin are described better on the basis of the Kramers equation.[90,107] Aleksandrov and Gol'danskii[108] have shown that the Arrhenius pre-exponential term may be considered a function of mobility in the reactant environment, and thus the transition state theory may be formally applied.

As a spectroscopist, it is interesting to analyse the nature of the high rate and specificity of enzyme reactions, or at least those reactions which include considerable changes in polarity but small changes of reactant configuration (e.g. proton transfer), suggesting an analogy between activated and electronic excited states. The activated transition state is the state in which the distribution of charge differs from those of both substrate and product ground states but corresponds to an equal weight of both of them. Thus, as in the case of an electronically excited state, the transition state is not at equilibrium with the surrounding protein groups, whose relaxational motions could occur at the time of, and therefore influence, the reaction.

Consider a simple illustrative model. The activation energy of the enzyme process (or one of its stages) depends on dipolar interactions in the activated

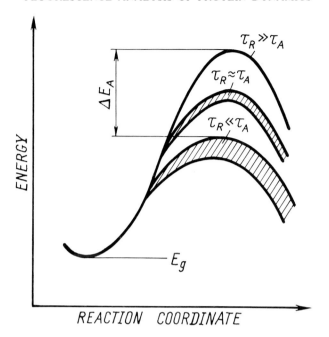

Fig. 9. Probable energy profiles for reactions with different relationships between dipolar relaxation time τ_R of the active site protein groups and effective lifetime of the transition intermediate τ_A. The cross-hatched area indicates the distribution in activation energy.

complex, while the mobility of protein dipoles is described by the relaxational time τ_R. Three cases with a different relationship between τ_R and the effective lifetime of the transition state complex τ_A may then be considered (Fig. 9). In the case that $\tau_R \gg \tau_A$, the orientations of the enzyme dipoles do not differ from those in the ground state. The specificity of substrate binding is highest, but the reaction proceeds slowly due to a high activation energy. The opposite limiting case is that of fast dipolar mobility, $\tau_R \ll \tau_A$. It is also expected to be unfavourable for the reaction, since the entropy of activation will be highest (rapid mobility makes correct fitting and orientation of substrate difficult) and there will probably be low specificity of binding. The intermediate case, $\tau_R \approx \tau_A$, i.e. when the mobility of the enzyme dipoles which surround the transition state complex is commensurate with the rate of an elementary stage of catalysis, is thought to be the most favourable for the reaction to proceed.

The real situation is probably more complex. Substrate binding occurs to a site exposed to solvent, while the reactions are thought to proceed in protein environments isolated from solvent.[103,109] Instead of a single energy barrier, multiple barriers are more typical of protein reactions. Again, the description of protein dynamics by a single relaxation time represents a possibly severe over-simplification. An interesting hierarchical model of protein states and motions has been presented recently.[110] It develops the concepts that protein motions of different magnitude occur on different time-scales,[29,111] and protein relaxations are sequential and non-exponential in time. Thus the mechanisms of specific binding and molecular recognition may be sequential.

Substrate or ligand binding produces a "protein-quake" which extends over the whole protein molecule[110] and may result in allosteric communication between different binding sites. Binding may cause changes in the frequencies and amplitudes of macromolecular thermal fluctuations in response to ligand binding and thus induce allostery without conformational change.[112]

VII. Outlook

In conclusion, we stress several problems related to biochemical mechanisms that may be solved with the aid of fluorescence spectroscopic studies of protein dynamics.

(1) Does the specific binding of substrates, inhibitors or cofactors induce a decrease in mobility in the binding sites at equilibrium? Does it result in an altered (more narrow) distribution of protein microstates? The data presented above indicate that fluorescence probe binding may result in immobilization of the binding sites in the nanosecond time-range, and the distribution of protein microstates in the binding sites may not differ substantially from that of chromophores in solid polar solvents. There is a need for extended studies of protein binding natural fluorescent protein ligands—substrates, inhibitors and cofactors and their modified fluorescent forms—by dynamic fluorescent depolarization and molecular relaxation spectroscopy. The spectroscopic properties of flavins and pyridine nucleotides allow such experiments to be performed. Several results on coenzyme motions obtained by time-resolved spectroscopy have been obtained already.[113-116]

(2) Does a correlation exist between the strength of ligand binding and its (or its environment's) immobilization on binding. It follows from general considerations that a decrease in the binding site mobility resulting in entropy reduction lowers the effective binding constant. However, comparative experimental data are lacking. Such studies could be performed, for

instance, on the binding of fluorescent haptens to antibodies of different specificity or hapten analogues with different affinity to antibodies of the same type. It would also appear interesting to observe the effect of binding on the fluorescence behaviour of tryptophans in the binding sites. The location of appropriate models for such studies is a pressing problem.

(3) If specific binding (recognition) means a decrease in mobility in the binding sites, can this decrease propagate over the whole protein molecule? Could it also be one of the mechanisms of allosteric regulation: allostery without "conformational change"? To obtain information on this problem one could measure the response of the spectral probe in one protein site to the binding of ligands at the other site.

(4) Are there any differences in the distribution of enzyme–substrate interaction energy in enzyme microstates between the ground and transition states? If the dispersion of microstates in the transition state is narrow, could it be an important mechanism determining the entropic effect of catalysis? Enzyme complexes with transition state analogues are currently under study by different methods including tryptophan fluorescence quenching.[117] There is a need for experiments directed towards determining the parameters of inhomogeneous broadening of spectra of tryptophans, both in active sites and as fluorescent transition state intermediates.

(5) To what extent are the elementary processes in enzyme catalysis and its regulation correlated with intramolecular motions in proteins? Fast time-scale molecular dynamics is an inevitable property of protein molecules considered as small thermodynamic systems. If not directly correlated with function they would be expected to hamper protein-catalysed reactions.[111] However, specific protein design, directed to a required function, may turn the inevitable to advantage. To clarify this important item, studies of structural dynamics in association with an analysis of the kinetic properties of proteins under the influence of different variables including temperature and solvent viscosity are necessary.

As observed above, direct fluorescence studies of protein dynamics are limited to the nanosecond time-range, and analysis of the distribution of microstates is possible in the absence of dynamic dipole relaxation in this time-range. This serious limitation may be overcome substantially by the use of low-temperature techniques to reduce protein mobility and bring it into the required longer time-range. Slowing down of dynamics is also expected in viscous media and in the membrane environment natural for many proteins. Thus the achievements and prospects of fluorescence spectroscopy make it useful for a range of basic structural and dynamic studies of biochemical mechanisms. It is of particular importance since, in this complicated area of research, no single method is decisive.

REFERENCES

1. Demchenko, A. P. (1986). *Ultraviolet Spectroscopy of Proteins*. Springer-Verlag, Berlin.
2. Lakowicz, J. R. (1983). *Introduction to Fluorescence Spectroscopy*. Plenum Press, New York and London.
3. Burstein, E. A. (1976). Luminescence of protein chromophores (model studies). *Ser. Biophysica*, Vol. 6, VINITI, Moscow.
4. Lumry, R. & Hershberger, M. (1978). Status of indole photochemistry with special reference to biological application. *Photochem. Photobiol.* **27**, 819–840.
5. Lakowicz, J. R. & Weber, G. (1973). Quenching of protein fluorescence by oxygen. Detection of structural fluctuations in proteins on the nanosecond time scale. *Biochemistry* **12**, 4171–4179.
6. Lakowicz, J. R. (1980). Fluorescence spectroscopic investigations of the dynamics properties of proteins, membranes and nucleic acids. *J. Biochem. Biophys. Methods.* **2**, 91–119.
7. Eftink, M. R., Ghiron, C. A. (1981). Fluorescence quenching studies with proteins. *Anal. Biochem.* **114**, 199–227.
8. Munro, J., Pecht, I. & Stryer, L. (1979). Subnanosecond motions of tryptophan residues in proteins. *Proc. Natl. Acad. Sci. USA* **76**, 56–60.
9. Lakowicz, J. R., Maliwal, B. P., Cherek, H. E. & Balter, A. (1983). Rotational freedom of tryptophan residues in proteins and peptides. *Biochemistry* **22**, 1741–1751.
10. Demchenko, A. P. (1985). Fluorescence molecular relaxation studies of protein dynamics. The probe binding site of melittin is rigid on nanosecond timescale. *FEBS Lett.* **182**, 99–102.
11. Demchenko, A. P. & Ladokhin, A. S. (1986). Red-edge-excitation fluorescence spectroscopy of indole and tryptophan. *Eur. Biophys. J.* (in press).
12. Frauenfelder, H., Petsko, G. A. & Tsernoglou, D. (1979). Temperature dependent X-ray diffraction as a probe of protein structure dynamics. *Nature (London)* **280**, 558–563.
13. Hartmann, H., Parak, F., Seigelmann, W., Petsko, G. A., Ponzi, D. R. & Frauenfelder, H. (1982). Conformational substates in a protein; Structure and dynamics of metmyoglobin at 80 K. *Proc. Natl. Acad. Sci. USA* **79**, 4967–4971.
14. Goldanskii, V. T., Krupyansky, Yu. F. & Frolov, E. N. (1983). Role of conformational sub-states in reaction capacity of protein molecules. *Mol. Biol. (USSR)* **17**, 532–542.
15. Gave, A., Dobson, C. M., Parello, J. & Williams, R. J. P. (1976). Conformational mobility within the structure of muscular parvalbumin. An NMR study of the aromatic resonances of phenylalanine residues. *FEBS Lett.* **65**, 190–194.
16. Williams, R. J. P. (1978). Energy states of proteins, enzymes and membranes. *Proc. Roy. Soc. London* **B200**, 353–389.
17. Fretto, L. & Strickland, E. H. (1971). Effect of temperature upon the conformations of carboxypeptidase A. *Biochim. Biophys. Acta* **235**, 473–488.
18. Demchenko, A. P. & Zyma, V. L. (1975). Thermal perturbation spectroscopy of proteins. I. Medium polarity effect. *Stud. Biophys.* **52**, 209–221.
19. Demchenko, A. P. & Zyma, V. L. (1977). Thermal perturbation spectroscopy of proteins. II. Origin of model chromophore spectra. *Stud. Biophys.* **64**, 143–150.

20. Hershberger, M. V., Maki, A. H. & Galley, W. C. (1970). Phosphorescence and optically detected magnetic resonance studies of a class of anomalous tryptophan residues in globular proteins. *Biochemistry* **19**, 2204–2209.
21. Friedman, J. M. & Scott, T. W. (1983). Hemoglobin dynamics. *Photochem. Photobiol.* **37**, 57.
22. Galley, W. C. & Purkey, R. M. (1970). Role of heterogeneity of the solvation site of in electronic spectra in solution. *Proc. Natl. Acad. Sci. USA* **67**, 1116–1121.
23. Rubinov, A. N. & Tomin, V. I. (1970). Bathochromic luminescence in low-temperature solutions of dyes. *Optics Spectr.* (*USSR*) **29**, 1082–1086.
24. Wuthrich, K. & Wagner, G. (1984). Internal dynamics of protein. *Trends Biochem. Sci.* **9**, 152–155.
25. Wittebort, R. J., Rothgeb, T. M., Szabo, A. & Gurd, F. (1979). Aliphatic groups of sperm whale myoglobin. ^{13}C-NMR study. *Proc. Natl. Acad. Sci. USA* **76**, 1059–1063.
26. Bystrov, V. F. (1984). Impact of nuclear magnetic resonance on structure-functional relationship study of proteins and peptides. *Bioorg. Chem.* (*USSR*) **10**, 997–1043.
27. Likhtenstein, G. I. & Kotelnikov, A. I. (1983). The study of fluctuational intramolecular lability of protein by physical labeling. *Mol. Biol.* (*USSR*) **17**, 505–518.
28. Domanus, I., Strambini, G. B. & Galley, W. C. (1980). Heterogeneity in the thermally-induced quenching of the phosphorescence of multi-tryptophan proteins. *Photochem. Photobiol.* **31**, 15–21.
29. Careri, G., Fasella, P. & Gratton, E. (1975). Statistical time events in enzymes CRC. *Crit. Rev. Biochem.* **3**, 141–164.
30. Careri, G., Fasella, P. & Gratton, E. (1979). Enzyme dynamics: the statistical physics approach. *Annu. Rev. Biophys. Bioeng.* **8**, 69–97.
31. Eftink, M. R. & Ghiron, C. A. (1977). Exposure of tryptophanyl residues and protein dynamics. *Biochemistry* **16**, 5546–5551.
32. Bushueva, T. L., Busel, E. P. & Burstein, E. A. (1980). Some regularities of dynamic accessibility of buried fluorescent residues to external quenchers in proteins. *Arch. Biochem. Biophys.* **204**, 161–166.
33. Miller, S. I. & Narutis, V. P. (1984). Hydrogen exchange and macromolecular mobility. *Biochemistry* **23**, 5113–5118.
34. Calhoun, D. B., Vanderkooi, J. M., Woodrow, G. V. & Englander, S. W. (1983). Penetration of dioxygen into proteins studied by quenching of phosphorescence and fluorescence. *Biochemistry* **22**, 1526–1533.
35. Calhoun, D. B., Vanderkooi, J. M. & Englander, S. W. (1983). Penetration of small molecules into proteins studied by quenching of phosphorescence and fluorescence. *Biochemistry* **22**, 1533–1540.
36. Bushueva, T. L., Busel, E. P. & Burstein, E. A. (1978). Relationship of thermal quenching of protein fluorescence to intramolecular structural mobility. *Biochem. Biophys. Acta* **534**, 141–152.
37. Burstein, E. A. (1977). Intrinsic luminescence of proteins (origin and applications). *Ser. Biophysica*, Vol. 7. VINITI, Moscow.
38. Burstein, E. A. (1983). Intrinsic protein luminescence as a tool for studying fast structural dynamics. *Mol. Biol.* (*USSR*) **17**, 455–467.
39. Wijnaendts van Resandt, R. W. W. (1983). Picosecond transient effect in the fluorescence quenching of tryptophan. *Chem. Phys. Lett.* **95**, 205–208.

40. Lakowicz, J. R. & Weber, G. (1980). Nanosecond segmental mobilities of tryptophan residues in proteins observed by lifetime-resolved fluorescence anisotropies. *Biophys. J.* **32**, 591–601.
41. Eftink, M. (1983). Quenching-resolved emission anisotropy studies with single and multitryptophan-containing proteins. *Biophys. J.* **43**, 323–334.
42. Turoverov, K. K. & Kuznetzova, I. M. (1983). Polarization of intrinsic fluorescence of proteins. II. Application for studies on equilibrium dynamics of tryptophan residues. *Mol. Biol. (USSR)* **17**, 468–474.
43. Rholam, M., Scarlata, S. & Weber, G. (1984). Frictional resistance to the local rotations of fluorophores in proteins. *Biochemistry* **23**, 6793–6796.
44. Lakowicz, J. R. (1984). Time-dependent rotational rates of excited fluorophores. A linkage between fluorescence depolarization and solvent relaxation. *Biophys. Chem.* **19**, 13–24.
45. Beddard, G. S. & Tran, C. D. (1985). Fluorescence studies of the restricted motion of tryptophan in α-cobre-toxin. *Eur. Biophys. J.* **11**, 243–248.
46. Takashima, S. (1969). Dielectric properties of proteins. I. Dielectric relaxation. In *Physical Principles and Techniques of Protein Chemistry* (Leach, S. J., ed.), Part A, pp. 291–333. Academic Press, New York and London.
47. Mazurenko, Yu. T. & Bakhshiev, N. G. (1970). Effect of the orientation dipolar relaxation on spectral, temporal and polarization characteristics of luminescence in solutions. *Optics Spectr. (USSR)* **28**, 905–913.
48. Bakhshiev, N. G. (1972). *Spectroscopy of Intermolecular Interactions.* Nauka, Leningrad, 263 pp.
49. Mazurenko, Yu. T. (1980). Dynamics of electronic spectra of solutions. Stohastic theory. Photoluminescence spectra. *Optics Spectr. (USSR)* **48**, 704–711.
50. Macgregor, R. B. & Weber, G. (1981). Fluorophores in polar media. Spectral effects of the Langevin distribution of electrostatic interactions. *Ann. N.Y. Acad. Sci.* **366**, 140–154.
51. Demchenko, A. P. (1982). On the nanosecond mobility in proteins. Edge excitation fluorescence red shift of protein-bound 2-(p-toluidinylnaphthalene)-6-sulfonate. *Biophys. Chem.* **15**, 101–109.
52. Samokish, V. A., Anufrieva, E. V. & Volkenstein, M. V. (1971). Spectroscopic investigation on α-chymotrypsin-proflavin interaction. *Mol. Biol.* **5**, 711–717.
53. Permyakov, E. A. & Burstein, E. A. (1975). Relaxation processes in frozen aqueous solution of proteins; temperature dependence of fluorescence parameters. *Stud. Biophys.* **51**, 91–103.
54. Eftink, M. R. & Ghiron, C. A. (1976). Exposure of tryptophanyl residues in proteins. Quantitative determination by fluorescence quenching studies. *Biochemistry* **15**, 672–680.
55. DeToma, R. P., Easter, J. H. & Brand, L. (1976). Dynamic interactions of fluorescence probes with the solvent environment. *J. Am. Chem. Soc.* **98**, 5001–5007.
56. Okamura, T., Sumitani, M. & Yoshihara, K. (1983). Picosecond dynamics Stokes shift of α-naphthylamine. *Chem. Phys. Lett.* **94**, 339–343.
57. Brand, L. & Gohlike, J. P. (1971). Nanosecond time-resolved fluorescence spectra of a protein–dye complex. *J. Biol. Chem.* **246**, 2317–2324.
58. Gafni, A., DeToma, R. P., Manrow, R. E. & Brand, L. (1977). Nanosecond decay studies of a fluorescence probe bound to apomyoglobin. *Biophys. J.* **17**, 155–168.

59. Lakowicz, J. R. & Cherek, H. (1981). Proof of nanosecond time-scale relaxation by phase fluorometry. *Biochem. Biophys. Res. Commun.* **99**, 1173–1178.
60. Grinvald, A. & Steinberg, I. Z. (1974). Fast relaxation processes in a protein revealed by the decay kinetics of tryptophan fluorescence. *Biochemistry* **13**, 5170–5177.
61. Grinvald, A. & Steinberg, I. Z. (1976). The fluorescence decay of tryptophan residues in native and denatured proteins. *Biochem. Biophys. Acta* **427**, 663–678.
62. Lakowicz, J. R. & Cherek, H. (1980). Dipolar relaxation in proteins on the nanosecond timescale observed by wavelength-resolved phase fluorometry of tryptophan residues. *J. Biol. Chem.* **255**, 831–834.
63. Lakowicz, J. R. & Keating-Nakamoto, S. (1984). Red-edge excitation of fluorescence and dynamic properties of proteins and membranes. *Biochemistry* **23**, 3013–3021.
64. Demchenko, A. P. (1981). Dependence of human serum albumin fluorescence spectrum on the excitation wavelength. *Ukr. Biochem. Z. (USSR)* **53** (3), 22–27.
65. Demchenko, A. P. (1986). Edge-excitation fluorescence spectroscopy of single-tryptophan proteins. *Eur. Biophys. J.* (in press).
66. Glushak, V. N., Demchenko, A. P., Orlovska, N. N. & Guly, M. F. (1984). Spectral characteristics of muscular aspartyl- and valyl-tRNA-synthetases and their complexes with substrates in norm and after long-term starvation. *Ukr. Biochem. Z. (USSR)* **56**, 519–526.
67. Demchenko, A. P. & Shcherbatska, N. V. (1985). Nanosecond dynamics of charged fluorescent probes at the polar interface of membrane phospholipid bilayers. *Biophys. Chem.* **22**, 141–143.
68. Demchenko, A. P. (1984). Structural relaxation in protein molecules studied by fluorescence spectroscopy. *J. Mol. Struct.* **114**, 45–48.
69. Perutz, M. F. (1980). Electrostatic effects in proteins. *Science* **201**, 1187–1191.
70. Wada, A. & Nakamura, H. (1981). Nature of the charge distribution in proteins. *Nature (London)* **293**, 757–758.
71. Rashin, A. A. & Honig, B. (1984). On the environment of ionizable groups in globular proteins. *J. Mol. Biol.* **173**, 515–521.
72. Sheridan, R. P. & Allen, L. C. (1980). The electrostatic potential of the alpha helix. *Biophys. Chem.* **11**, 133–136.
73. Holl, W. G. J., Halic, L. M. & Sander, C. (1981). Dipoles of the α-helix and β-sheet: their role in protein folding. *Nature (London)* **294**, 532–536.
74. Rees, D. C. (1980). Experimental evaluation of the effective dielectric constant of proteins. *J. Mol. Biol.* **141**, 323–326.
75. Honig, B., Gilson, M., Fine, R. & Rashin, A. (1984). Calculation of the effective dielectric constant in the interior of proteins. *Biophys. J.* **45**, 129a.
76. Warshel, A., Russell, S. T. & Churg, A. K. (1984). Macroscopic models for studies of electrostatic interactions in proteins: limitations and applicability. *Proc. Natl. Acad. Sci. USA* **81**, 4785–4789.
77. Kuntz, I. D. & Kauzmann, W. (1974). Hydration of proteins and polypeptides. *Adv. Protein Chem.* **28**, 239–345.
78. Rupley, J. A., Gratton, E. & Careri, G. (1983). Water and globular proteins. *Trends Biochem. Sci.* **8**, 18–22.
79. Robinson, G. W., Robinson, R. J., Fleming, G. R., Morris, J. M., Knight, A. E. W. & Morrison, R. J. S. (1978). Picosecond studies of the fluorescence

probe molecule 8-anilino-I-naphthalene sulfonic acid. *J. Am. Chem. Soc.* **100**, 7145–7150.

80. Landau, L. D. & Lifshitz, E. M. (1958). *Statistical Physics*. Pergamon Press, London.

81. Weiner, P., Langridge, R., Blaney, J., Schaefer, K. & Kollman, P. A. (1982). Electrostatic potential molecular surfaces. *Proc. Natl. Acad. Sci. USA* **79**, 3754–3758.

82. Van Duijen, P. T., Thole, B. T. & Holl, W. G. (1979). On the role of the active site helix in papain, *ab initio* molecular orbital study. *Biophys. Chem.* **9**, 273–280.

83. Van Duijen, P. T., Thole, B. T., Broer, R. & Nieuwpoort, W. C. (1980). Active-site α-helix in papain and the stability of the ion pair RS⁻ . . . Im H⁺. *ab initio* molecular orbital study. *Int. J. Quantum Chem.* **17**, 651–671.

84. Warwicher, J. & Watson, H. C. (1982). Calculation of the electric potential in the active site cleft due to α-helix dipoles. *J. Mol. Biol.* **157**, 671–679.

85. Koppenol, W. H. (1980). Effect of a molecular dipole on the ionic strength dependence of a bimolecular rate constant. Identification of the site of reaction. *Biophys. J.* **29**, 493–507.

86. Warshel, A. (1978). Energetics of enzyme catalysis. *Proc. Natl. Acad. Sci. USA* **75**, 5250–5254.

87. Krishtalik, L. I. (1980). Catalytic acceleration of reactions by enzymes. Effect of screening of a polar medium by a protein globule. *J. Theor. Biol.* **86**, 757–772.

88. Beece, D., Eisenstein, L., Frauenfelder, H., Good, D., Marden, M. C., Reinisch, L., Reynolds, A. H., Sorensen, L. B. & Yue, K. T. (1980). Solvent viscosity and protein dynamics. *Biochemistry* **19**, 5147–5157.

89. Alberding, N., Chan, S. S., Eisenstein, L., Frauenfelder, H., Good, D., Gunsalus, I. S., Nordlund, T. M., Perutz, F. M., Reynolds, A. H. & Sorensen, L. B. (1978). Binding of carbon monoxide to isolated hemoglobin chains. *Biochemistry* **17**, 43–51.

90. Debrunner, P. G. & Frauenfelder, H. (1982). Dynamics of proteins. *Annu. Rev. Phys. Chem.* **33**, 283–299.

91. Läuger, P. (1985). Ionic channels with conformational substates. *Biophys. J.* **47**, 581–590.

92. Lamola, A. A. & Flores, J. (1982). Effect of buffer viscosity on the fluorescence of bilirubin bound to human serum albumin. *J. Am. Chem. Soc.* **104**, 2530–2534.

93. Gavish, B. & Werber, M. M. (1979). Viscosity-dependent structural fluctuations in enzyme catalysis. *Biochemistry* **18**, 1269–1275.

94. Whitternburg, S. L. (1983). Microviscosity and enzyme–ligand dissociation. *J. Theor. Biol.* **101**, 675–683.

95. Gorodysky, V. A. (1984). The influence of orientational non-equilibrium solvatition of transition state on simple equilibrium in solutions. *Z. Phys. Chim. (USSR)* **58**, 2485–2488.

96. Van der Zwan, G. & Hynes, J. T. (1983). Polarization diffusion effects on reaction rates in polar solvents. *Chem. Phys. Lett.* **101**, 367–371.

97. Krishtalik, L. I. & Kharkats, Yu. I. (1984). Reorganization energies of the medium in the charge transfer reaction in the protein globule. *Biophysica.* **29**, 19–22.

98. Dogonadze, R. R., Kuznetsov, A. M. & Marsagishvili, T. A. (1980). The present state of the theory of charge transfer processes in condensed phase. *Electrochimica Acta* **25**, 1–28.

99. Northrup, S. H., Zarrin, F. & McCammon, J. A. (1982). Rate theory for gated diffusion-influenced ligand binding to proteins. *J. Phys. Chem.* **86**, 2314–2321.

100. Northrup, S. H., Rear, M. R., Lee, C.-Y., McCammon, J. A. & Karplus, M. (1982). Dynamical theory of activated processes in globular proteins. *Proc. Natl. Acad. Sci. USA* **79**, 4035–4039.

101. Shaitan, K. V. & Rubin, A. B. (1983). Bending fluctuations of α-helices and dynamics of enzyme–substrate interactions. *Mol. Biol. (USSR)* **17**, 1280–1296.

102. Welch, G. R., Somogyi, B. & Damjanovich, S. (1982). The role of protein fluctuations in enzyme: A review. *Progr. Biophys. Mol. Biol.* **39**, 109–146.

103. Somogyi, B., Welch, G. R. & Damjanovich, S. (1984). The dynamic basis of energy transduction in enzymes. *Biochem. Biophys. Acta* **768**, 81–112.

104. Schlitter, J. (1984). Rapid energy exchange of enzymes with solvent water by dipolar interactions. *J. Theor. Biol.* **106**, 303–313.

105. Warshel, A. (1984). Dynamics of enzymatic reactions. *Proc. Natl. Acad. Sci. USA* **81**, 444–448.

106. Dogonadze, R. R., Kuznetsov, A. M. & Ulstrup, J. (1977). Conformational dynamics in biological electron and atom transfer reactions. *J. Theor. Biol.* **69**, 239–263.

107. Doster, W. (1983). Viscosity scaling and protein dynamics. *Biophys. Chem.* **17**, 97–103.

108. Aleksandrov, I. V. & Gol'danskii, V. I. (1984). On the dependence of the rate constant of an elementary chemical process in the kinetic region on the viscosity of the medium. *Chem. Phys.* **87**, 455–466.

109. Dewar, M. J. S. & Storch, D. M. (1985). Alternative view of enzyme reactions. *Proc. Natl. Acad. Sci. USA* **82**, 2225–2229.

110. Ausari, A., Berendzen, J., Bowne, S. F., Frauenfelder, H., Iben, I. E. T., Sauke, T. B., Shyamsunder, E. & Young, R. D. (1985). Protein states and proteinquakes. *Proc. Natl. Acad. Sci. USA* **82**, 5000–5004.

111. Demchenko, A. P. (1981). Equilibrium intramolecular mobility in proteins. *Ukr. Biochim. Z.* **53**, 114–128.

112. Cooper, A. & Dryden, D. T. F. (1984). Allostery without conformational change, A plausible model. *Eur. Biophys. J.* **11**, 103–109.

113. Gafni, A. & Brand, L. (1976). Fluorescence decay studies of reduced nicotinamide adenine dinucleotide in solution and bound to liver alcohol dehydrogenase. *Biochemistry* **15**, 3165–3171.

114. Brochon, J.-C., Wahl, Ph., Monneuse-Doublet, M.-O. & Olomucki, A. (1977). Pulse fluorimetry study of octopine dehydrogenase reduced nicotinamide adenine dinucleotide complexes. *Biochemistry* **16**, 4594–4599.

115. Visser, A. J. W. G., Grande, H. J. & Veeger, C. (1980). Rapid relaxation processes in pig heart lipoamide dehydrogenase revealed by subnanosecond resolved fluorometry. *Biophys. Chem.* **12**, 35–49.

116. Visser, A. J. W. G. (1986). Excited states of flavins. In *Excited State Probes in Biochemistry and Biology* (Szabo, A. G. & Masotti, L., eds). Plenum Press, New York and London.

117. Kurz, L. C., Lazard, D. & Frieden, C. (1985). Protein structural changes accompanying formation of enzymatic transition states: tryptophan environment in ground-state and transition-state analogue complexes of adenosine deaminase. *Biochemistry* **24**, 1342–1346.

The Origin and Use of the Terms Competitive and Non-competitive in Interactions among Chemical Substances in Biological Systems

HENRY McILWAIN

Department of Biochemistry, St Thomas's Hospital Medical School
(United Medical and Dental Schools, London University),
Lambeth Palace Road, London SE1 7EH, England

I. Foreword 158
II. Establishing the "Competitive" Terminology 159
   A. The First Examples 159
   B. O_2/CO Interactions in Man and the Mouse 161
   C. O_2/CO Interactions in a Moth and a Plant 162
   D. Invertase, β-D-fructofuranosidase 162
III. Chemical Usage of "Competitive" 165
IV. Applications in Enzyme Studies 166
   A. Hydrolases 167
   B. Oxidoreduction 168
   C. Further Analyses of Inhibitory Type 169
V. Applications in Microbiology 171
VI. Applications in Toxicology and Pharmacology 174
   A. Drug Antagonism 174
   B. Sulphonamides and Competition 176
   C. Binding Studies, Receptors and Competition . . . 177
VII. Competitive Terminology: Technical and Non-technical . 179
VIII. Summary 181
   Acknowledgements 182
   References 182

I. Foreword

The introduction and use of the terms "competitive" and "non-competitive" reflect in an interesting way the growth of biochemical, physiological and pharmacological knowledge during the past 100 years. Use of the terms has been the subject of many past and recent misstatements, as documented below. It is intriguing that in enzyme studies the distinction was made between phenomena later called competitive and non-competitive inhibition, and that the distinction was given a firm experimental basis, some 20 years before those names were allocated.

ESSAYS IN BIOCHEMISTRY Vol. 22
ISBN 0 12 158122 5

According to the Oxford English Dictionary,[1] the word competition appeared in English in the seventeenth century and meant the action of attempting to gain what another tries to gain at the same time; the striving of two or more for the same objective. It reached less personal applications in describing social policy, Darwinism and the law of mass action, in the nineteenth century (see below). In its current ordinary and scientific uses, I judge that the word competition implies some initial similarity in status of the competing entities, and subsequently a time-process which affords a rank order or proportionation among them. In competition, such allocation must remain open to readjustment for at least a certain period. In many ordinary uses, uncompetitive or co-operative are antonyms to competitive, but this is not usually the case in biochemistry (see, however, Section IV.C).

II. Establishing the "Competitive" Terminology

A. THE FIRST EXAMPLES

Application of the "competitive" terminology for interactions among chemical substances in biology appears to have entered English usage from its German equivalent. The first comprehensive use which I have found of the term "competitive inhibition", in its present sense, is by J. B. S. Haldane[2] in 1927, in his account of carbon monoxide as a tissue poison. This paper includes confirmation and extension of some of the work of Otto Warburg[3] of 1926 in which Warburg observed that the concentration of carbon monoxide needed to inhibit respiration of a yeast and of a bacterial preparation depended on the concomitant concentration of oxygen (Fig. 1a).

This investigation was part of Warburg's study of cell respiration which he was already interpreting in enzymic terms with *das Atmungsferment* as the responsible agent; resolution of the complexities of the respiratory chain was just commencing in his Berlin laboratory, and in Cambridge by Keilin and by others including Haldane. In the majority of instances when Warburg described the interaction of O_2 and CO the expression used is one such as

> . . . das Atmungsferment verteilt sich zwischen beiden Gasen. Deshalb hemmt ein bestimmter Kohlenoxyd Druck die Atmung so starker, je niedriger der Sauerstoffdruck.

CO inhibited more strongly, the lower the O_2 pressure. However, in one instance the description is significantly different:

> Hierbei bemühte ich mich, den Sauerstoff druck so tief als möglich herabzu-

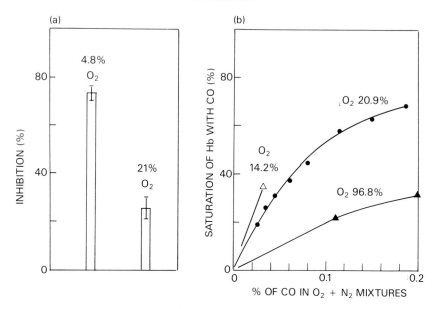

Fig. 1. Increasing oxygen concentration opposing the actions of carbon monoxide:
(a) in inhibiting the respiration of yeast (Warburg[3]). The inhibitory action of 79% v/v of CO is shown in the presence of two oxygen concentrations (v/v); other component, N_2.
(b) in depressing the formation of carboxyhaemoglobin in people breathing air or O_2–N_2 mixtures with the specified concentrations of CO (Haldane[12]).

drucken, weil nach den oben mitgeteilten Erfahrungen an eine Konkurrenz zwischen Sauerstoff und Kohlenoxyd zu denken war.

That is, "a competition was to be thought of between O_2 and CO".

Eine Konkurrenz appears fully equivalent in its German usage to the English "competition", being used in sport, or for a rival shop, and *Konkurrenz Prüfung* for competitive examination. It overlapped in meaning with *der Wettbewerb*, competitive prices being *wettbewerbsfahig*. However, *Konkurrentz* had[4] a relevant physiological use in describing reflexes competing with effects of external stimulation.

In a later account of his work in book form in which the O_2/CO competition for the *Atmungsferment* is again given prominence and which is based on the same experimental data, Warburg[5] used neither the term *Konkurrenz* nor *Wettbewerb*, but rather the expressions noted above for the enzyme being partitioned between O_2 and CO, or the oxygen being "displaced" (see below). J. B. S. Haldane, however, not only generalized the term in 1927 but gave prominence to its use in his monograph *Enzymes*[6] of 1930. Unless earlier comparable instances are uncovered I judge that

Haldane is to be regarded as the originator of the current description of specified types of inhibition, in metabolic and enzyme studies, as competitive or non-competitive. In neither his 1927 nor his 1930 publication does Haldane refer to the terms as personal introductions, nor indeed does Warburg,[3] write similarly of *die Konkurrenz*, Warburg's[3] sentence quoted above is, however, written in the first person and *"bemühte ich mich"* and *"zu denken war"* give to the use of *eine Konkurrenz* the flavour of a personal introduction to the use of competitive relationships; but it will be noted that he did not introduce a term comparable to non-competitive, nor did he continue his use of "competitive".

In generalizing the use of "competitive" nomenclature Haldane[2] included: (1) Warburg's[3] experiments; (2) O_2/CO interactions with haemoglobin in man; (3) an O_2/CO interaction in moths, the investigation of which was suggested by Keilin; and (4) observations with yeast saccharase or invertase. Items (2) to (4) will now be considered separately.

B. O_2/CO INTERACTIONS IN MAN AND THE MOUSE

J. B. S. Haldane's[2] interest in Warburg's[2] paper can readily be understood: at the age of 19 or 20 J.B.S. was one of two co-authors with his father J. S. Haldane, of a paper on CO/haemoglobin interactions in man.[7] Moreover, joint work of his father's[8] in 1896 had with great perspicacity observed the effect of illumination in dissociating carboxyhaemoglobin, a discovery of which Warburg[3] made admirable use in investigating tissue metabolism. This work had originated in a noteworthy application of chemistry and physiology to practical matters in which J. S. Haldane played a major part: the investigation of toxic mine gases.[9-11] J. B. S. Haldane, also, had other interests which used "competitive" terminology (Section VII).

Data from Haldane & Lorrain Smith's[8] 1896 paper, given in Fig. 1b suggest relationships which were later termed competitive, though that term is not used in their account. Combination of haemoglobin with a given concentration of CO is reversible and is diminished by increased concentrations of O_2; "the proportions in which it divides itself between the two gases depending, in accordance with the law of mass action, on the relative partial pressures of the two gases".[7] Thus haemoglobin was equally divided between O_2 and CO when exposed to air conditioning about 0·07% of CO, v/v, and Haldane[11] advocated the breathing of O_2 (with some CO_2) in place of air, in treatment of carbon monoxide poisoning: "it increases greatly the amount of dissolved oxygen in the blood . . . and expels far more rapidly the CO of CO haemoglobin". Suggestive data, but not the terminology, of competitive relationships were thus evident in the 1922 writing.

Interacting concentrations of CO and O_2 differ greatly in the observations of Figs 1a and b: CO has much greater affinity for haemoglobin than for Warburg's respiratory enzyme. As noted by Warburg,[5] J. S. Haldane[12] in 1895 had shown that mice remained alive in a gas mixture of 1 atmosphere pressure of CO plus 2 atmospheres of O_2, conditions under which almost all the animal's haemoglobin is present as carboxyhaemoglobin, but when the O_2 dissolved in the blood is sufficient for tissue respiration. Haldane[12] thus concluded that apart from its action at haemoglobin, CO was as physiologically indifferent as N_2. In this experiment, however, the hyperbaric O_2 was displacing CO from its combination at tissue level, for J. B. S. Haldane[2] subsequently confirmed J. S. Haldane's[12] experiments but examined also the effects of 1 atmosphere of CO with greater CO/O_2 ratios; these affected movements of the animals and later caused convulsions, an action attributed by J. B. S. Haldane to CO competing with O_2 for a tissue respiratory component, probably "an iron compound analogous to haemoglobin" for which, however, CO had much less affinity.

C. O_2/CO INTERACTIONS IN A MOTH AND A PLANT

J. B. S. Haldane's[2] findings on the susceptibilities of wax moths to CO are shown in Fig. 2a. The moths are much less susceptible to CO than are mammals; nevertheless competing relationships with O_2 are shown and are named as such. Linear relationships between [CO] and the relieving [O_2] were found over a large part of the range of concentrations examined, but the departure from linearity at lower concentrations was significant and was attributed to the presence of more than one critical CO-sensitive system. On returning the CO-immobilized moths to air, they regained activity. When the high concentrations of CO needed for inhibiting moth mobility were replaced by N_2, corresponding inhibition was not observed.

The high CO concentrations of Fig. 2a inhibited not only the yeast and a coccus as reported by Warburg,[3] but also the germination of cress seeds (*Lepidium sativum*); greater concentrations of CO were required the higher the concomitant oxygen tension.

D. INVERTASE, β-D-FRUCTOFURANOSIDASE

The enzyme EC 3.2.1.2.6, known by the names invertin, sucrase and saccharase, was extensively investigated from the 1870s onwards (for early literature see refs 13, 14). It was the first extracted and purified enzyme in which the relationships later described as competitive and non-competitive were differentiated, this being achieved in 1913 or earlier. It was also the first enzyme defined in terms of substrate and reaction catalysed, to which the

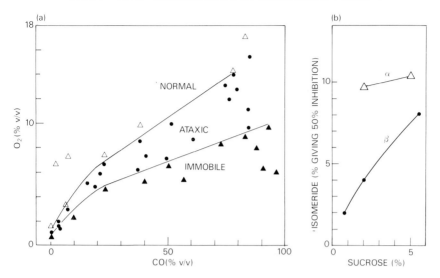

Fig. 2. (a) Reversibly inhibited motility of the wax moth, *Galleria mellonella*, by carbon monoxide in N_2–O_2–CO mixtures at 20°C. Moths were exposed to the mixtures, their degree of activity recorded, and the mixtures subsequently analysed: their $[O_2]$ and $[CO]$ giving coordinates for the points of the figure. Lines were drawn at the onset of ataxia and immobility (ref. 2). (b) Increase in concentration of glucose isomerides needed to give 50% inhibition of sucrose hydrolysis by a yeast invertase preparation as sucrose concentration is increased (based on data from ref. 24).

competitive terminology was applied: for Warburg's[3] prior use of the term was to the *Atmungsferment* which was not then so defined.

Invertase was the prototypical example of the extractable "unorganized ferments" for which Kuhne[15] in 1878 coined the term "enzyme"; it has been subjected to purification by adsorption and precipitation.[13,16] Such preparations of invertase and of two other enzymes were studied by Fischer[17] in the structural chemical and stereochemical work in which his famous lock and key simile was offered for the interpretation of enzyme specificity. Fischer's work on carbohydrates was initially in synthetic organic chemistry, but increasingly the glucosides produced were examined as enzyme substrates (for example, with Armstrong in 1901[18]). By 1900 the course of invertase action was being interpreted in terms of mass action,[16] initial velocities being proportional to concentrations of enzyme and substrate within suitable concentration ranges; deviations from proportionality were noted and were interpreted in terms of inhibition by the products, glucose and fructose. Also, several of the synthetic glucosides proved inhibitory. Combination of enzyme and substrate was postulated to explain these phenomena and this was supported by other observations including the stabilizing effect of sucrose on invertase exposed to heat inactivation.[13] It may be noted that Woolley[19] in subsequently giving an account of the "discovery and development of ideas about antimetabolites", wrongly states that the early investigations of "carbohydrate-hydrolysing enzymes . . . are limited to inhibitions caused by a product of the reaction", whereas the inhibitory synthetic glucosides were substrate analogues which Woolley[19] considered were first examined only some 25 years after Fischer & Armstrong's work.[18]

Although complexities arose from the isomerization of the products, by their separate addition both glucose and fructose added as such were shown to inhibit invertase. These and other inhibitory actions on the enzyme were examined by Michaelis & Pechstein[20] during and following the study of the kinetics of invertase action which gave the often-quoted Michaelis & Menten[21] formula of 1913. Michaelis & Menten[21] had developed earlier work to show that the enzyme E reacted with sucrose S to yield glucose G and fructose F according to the equations:

$$E + S \rightleftharpoons E.S; \qquad E.S \rightarrow E + F + G$$

and that the velocity of the first, reversible stage could be rate-limiting and follow mass-action kinetics. With relatively high concentrations of products, however, it appeared that compounds $E.F$ and $E.G$ needed to be taken into account.

Substances other than the products also interacted with invertase, as shown by their effects on the progress of the reaction; examples were glycerol and mannitol, but not maltose nor lactose. Moreover, kinetic analysis of sucrose hydrolysis showed that added compounds displayed different types of inhibition; these were recognized also in yeast maltase (EC 3.2.1.20, α-D-glucosidase).[22] Some inhibitory actions were irreversible; of the reversible inhibitions of invertase, two categories were differentiated. The first was exemplified by α-methylglucosides and by glycerol, which inhibited to a degree almost independent of the concomitant concentration of sucrose. Inhibition by the second, exemplified by fructose, was markedly dependent on the concentration of sucrose present in the reaction mixture (Fig. 2b). Thus 45 mM-fructose inhibited the hydrolysis of 50 mM-sucrose by 60% while it inhibited that of 500 mM-sucrose by only 25%.

Michaelis and his co-workers[20,21] drew attention to the different categories of inhibition in the subtitles or titles of their papers: "On the nature of differing types of inhibition of enzymes"; "On the differing types of inhibition of invertase action". They offered an appropriate mechanism for the kinetic differences observed. Thus glycerol could be regarded as sequestering a portion of the enzyme, so lowering reaction velocity though not affecting the affinity of the remaining enzyme for its substrate. Fructose interacted at the catalytic process itself, lowering the affinity of the enzyme for its substrate as measured by the increased concentration of sucrose needed for a given rate of reaction, though the maximum velocity with higher substrate concentrations was unaffected. This was referred to as "a distribution of the enzyme between the two substances" (". . . *so tritt ein Verteilung der Ferment . . .*").

Michaelis and his co-workers had thus given in 1913–14 satisfactory data which enabled Haldane[2] to quote their work as exemplifying competitive and non-competitive inhibition of enzyme actions. Haldane[2] also specifically connected that work with the actions of Sections II.B and II.C, writing that "CO is clearly comparable with fructose rather than glycerol" in that it competed with O_2. I have not found "competitive" terminology used to describe enzyme–substrate–inhibitor relationships in pre-1927 literature. Thus it is not used by Armstrong,[23] who continued in London the investigations of carbohydrate-splitting enzymes which he had commenced with Fischer (see Section III for Armstrong's different use of "competition"). Armstrong's[23] description of maltase is that

> β-methylglucoside acts to retard the hydrolysis of maltose . . . part of the enzyme must combine with the glucoside and so be withdrawn from action . . . combine with the β-methylglucoside though quite unable to hydrolyse it.

Similarly, "competitive" is not used by Nelson & Anderson[24] in their study of invertase that afforded the data of Fig. 2b, which they interpreted in terms of the differing affinities for invertase shown by sucrose and the glucose and fructose isomerides. In some cases, it has been necessary also to take into account the diminution of water concentration occasioned by the high sucrose concentrations employed; for recent discussion of such mixed inhibition and of non-competitive inhibition, see refs 16, 25.

III. Chemical Usage of "Competitive"

Ideas approaching the "competitive" of Section II are inherent in descriptions of chemical affinity and of chemical kinetics which were the basis of Michaelis' enzyme kinetics; they were expressed in a variety of terms, often anthropomorphic. Different strivings or affinities of atoms resulted in proportionation of a base between two acids; coefficients of affinity were obtained by measurement of the rate of chemical change: ". . . when equilibrium is established by the competition of two acids for the same base . . ." (from Pattison Muir's[26] description in 1884 of Guldberg & Waage's and Ostwald's work of the 1860s).

Such is the background to Armstrong's[23] use of the word "competition" in his discussion of the hydrolysis of disaccharides, in which he compares the actions of acids and enzymes in this regard:

> whereas there is competition between the solvent water and the carbohydrate for the acid, water has very little attraction for the enzyme: in consequence, practically the whole of the enzyme present is taking part in the operation of hydrolysis.

In this account, hydrolysis was visualized as proceeding through oxonium compounds, using residual valencies of oxygen at the site of hydrolysis. Competition for the enzyme is thus suggested to occur between the other two reactants in the hydrolysis, the substrate and the water. This use of the term competition does not appear to have been pursued by Armstrong, nor to have been applied to relationships between substrate and inhibitors. Reference to Armstrong's[23] book is given in Haldane's[6] *Enzymes*, which discusses the enzymes investigated by Armstrong but not his use of "competition"; use of "non-competitive" has not been found in Armstrong's writings. Armstrong's[23] monograph is, however, noteworthy as a precursor to Haldane's[2,6] much more specific and theory-bound use of the competitive terminology.

Competing reactions remain a theme in organic chemistry: Robinson's[27] discussion of reactivity in terms of electron availabilities at different sites in a molecule refers to "atoms which compete successfully with alkyl groups" for control of substitution. Remick's[28] account of electronic interpretations in organic chemistry carries a section on competitive reactions and includes comments on substituents which compete in influencing rates of bromination. "Non-competitive" has not been encountered in this connection.

Studies of surface phenomena also report competition for space at the surface of aqueous solutions; Adam[29] includes "competition between molecules coming to the surface from the interior and insoluble molecules at the surface". Surface potentials on butyl alcohol solutions are smaller than on water, "observations pointing to a simple competition between butyl alcohol and insoluble substances".

In the examples of this section "compete" remains closer to its ordinary English usage, and not linked to a specific theoretical background as it is in Haldane's[2,6] categorization of enzyme inhibitions as competitive or non-competitive. In scientific use generally, however, "compete", has been chosen rather than its near synonyms contend, contest, or rival. Antagonize and displace also can claim scientific adoption (Sections II.A, VI.C) but for uses distinct from that of "compete".

IV. Applications in Enzyme Studies

A common factor likely to have contributed to Haldane's[2] bringing together the phenomena of CO toxicity and carbohydrate hydrolyses, was the mathematical treatment of the data. Haldane's prowess in applying mathematics to diverse biological problems has been noted.[30] His contribution to the account of haemoglobin[7] was the mathematical analysis of its interaction with O_2 and CO; also, prior to Haldane's extension of Warburg's[3] work, he had participated in a reanalysis of Michaelis & Menten's[21] enzyme kinetics which was based on invertase[31] and he was thus well placed for emphasizing a common factor in these otherwise disparate aspects of metabolism.

Haldane's subsequent monograph entitled *Enzymes*[6] also emphasized aspects of the subject susceptible to quantitative expression. The book of some 200 pages carried over 600 references, and was written at a time when accounts of enzymes were some five to ten times longer than that. Haldane's book was thus concerned with principles: giving evidence for union of enzyme with substrates and related compounds; the theory and classification of enzyme action; effects of temperature, pH and toxic agents, with some details of their purification and specificity. Enzyme kinetics and Michaelis' theory are presented early, with some 35 pages of exposition, and with Michaelis constants quoted in tables and on some 20 other occasions throughout the text. Moreover, the competitive/non-competitive nomenclature is prominently displayed, with some 27 index entries. Many of these relate to the carbohydrate-splitting enzymes referred to above, but include some additional aspects of novel orientation, as in the use of a range of substrates or inhibitors giving quantitative data to establish the individuality and purity of enzymes during fractionation.

The terms "competitive" and "non-competitive" were thus introduced together with considerable theoretical background and with the convenient Michaelis constant expressing concentrations for 50% action. Haldane[6] regretted the frequent absence from published work of data adequate to differentiate between competitive and non-competitive interactions at a given enzyme, but was able to suggest the occurrence of such inhibition in several additional instances, which are included in Sections IV.A and B.

A. HYDROLASES

Data on esterases and lipases from numerous sources are collated in Haldane's[6] monograph, frequently indicating the substrate concentration for half maximum velocity to be a few mM. The investigation of lipases by Murray[32] acknowledged help and advice from Haldane, demonstrated competitive relationships with substrate analogues, and used "competitive" nomenclature. In it, hydrolysis of ethyl butyrate was measured manometrically using a lipase preparation made by aqueous extraction of an acetone powder from fat-free pig pancreas.

Ketones and aldehydes inhibited the hydrolysis to a much greater extent than did other structurally related compounds, and of them acetophenone was especially potent. Its inhibition was overcome by increasing substrate concentration; at a butyrate/acetophenone ratio of 2·5, inhibition was 70% while at a ratio of 10, inhibition fell to 25%.

> When additional substrate was given to the system containing a certain quantity of inhibitor, a rise in velocity was obtained . . . showing that the inhibitor did not act through any deleterious effect on the enzyme but merely as a competitor with the substrate.

The degree of inhibition was found to be independent of moderate changes in the quantity of enzyme preparation present. A corresponding ether, anisole, was much less active; methylpropylketone, benzaldehyde and ethylbenzoate acted in the fashion of acetophenone, but acetophenone oxime was inactive as an inhibitor. These competitive inhibitions were contrasted with inhibition of the same enzyme preparation by fluoride, which was described as non-competitive: further addition of substrate did not increase the velocity of hydrolysis which had been depressed by an added fluoride.

An examination of the action of fatty acids on pepsin and urease[33] is quoted by Haldane[6] as offering evidence for partly competitive inhibition. Thus the oleic acid inhibition of urease was slightly relieved by increase in urea concentration; this is discussed by Velluz[33] in terms of micellar properties without reference to Michaelis' work or to competition. Actions of dipeptidases from yeast and kidney as reported by Grassman & Klenk[34] used analysis by methods of Michaelis & Menten (ref. 21, though wrongly quoted) and afforded to Haldane[6] an instance of competitive inhibition. This was the action of alanine in inhibiting hydrolysis of leucylglycine (though not of glycylglycine) to a degree relieved by increased substrate concentration.

B. OXIDOREDUCTION

Competitive relationships at oxidoreductases were first shown using washed bacterial suspensions and methylene blue as acceptor, by a technique developed by Harden & Zilva[35] and Thunberg (see ref. 36), which involved measuring the time taken to convert a given quantity of the blue to its colourless reduced form in the presence of potential H-donors. Succinic acid was an effective donor when *B. coli* (now *Escherichia coli*) was used, but the bleaching was found by Quastel & Wooldridge[37] to be inhibited by a few compounds including malonic acid: transfer from 7 mM succinic acid was 95% inhibited by 77 mM malonic acid. Also, inhibition of the bleaching in the presence of 1.4 mM malonic acid was shown to be increasingly alleviated as the concentration of succinate was increased from 1·4 to 70 mM. This interaction was not, however, described as competitive but was related to a theory of adsorption and active centres which were in certain ways differentiated from enzymes.[37,38] A similar relationship was shown and conclusions drawn regarding hydrogen transfer from lactic acid and its inhibition by oxalic acid. No reference was made in the papers to the work of Michaelis and co-workers[20,21] nor to their differentiation among types of inhibition, nor to Haldane's[2] description of these types as competitive and non-competitive. Thus at this stage Quastel & Wooldridge[37,38] preferred to interpret their results independently.

Webb[39] gave to Cook[40] the credit for first describing the malonate inhibition of succinate dehydrogenation (by succinate dehydrogenase, EC 1.3.99.1) as competitive. Cook's[40] paper did not, however, include experimental data to establish the point, but acknowledged help from J. B. S. Haldane. These considerations thus leave Haldane's[6] monograph as the primary source for recognition of the competitive nature of the malonate inhibition of succinate dehydrogenation. Misinterpretation with respect to malonate inhibition has also extended to regarding it as the original instance of competitive relationships in enzyme studies and as "the classical example"[41] of competitive inhibition; it has led also to incorrect personal ascriptions of data and concepts.[42-44]

C. FURTHER ANALYSES OF INHIBITORY TYPE

Criteria for competitive and non-competitive enzyme inhibition were clearly expressed by Haldane,[6] but he found few examples to which the criteria could be applied with precision, and instead he accepted any degree of lesser inhibition with increased substrate concentration, as an indication of an "at least partly competitive mechanism". Shortly afterwards, however, his colleague Woolf[45] (and see ref. 41), and also Lineweaver & Burk[46] devised the double reciprocal plots which gave graphical indication of inhibitory type and which gained widespread use in enzyme investigation (for these methods, and their successors, see refs 16, 25, 47). It may be judged that Haldane's own writing was a major factor in alerting investigators to the value of gathering data from which the type of inhibition could be specified in these ways. The occasions for analysis of inhibitory type greatly increased when, as recounted in Sections V and VI, competitive studies found great application in microbiology and pharmacology as well as in biochemistry itself.

Continuing here the more purely enzymic considerations, recognition grew of the value of establishing accurately that specified interactions were "fully competitive", corresponding to combination of S or of I at the same site on E. The new graphical methods gave a clear contrast between competitive and non-competitive relationships in the sense adumbrated in Section II.D. In addition, they confirmed the existence of "partially competitive" relationships when S and I combined with E at sites adjacent and capable of interacting, to form a complex $E.I.S.$ In certain instances this arose only from $E.S$ and I.[41,48,49] The inhibition was then described as "uncompetitive"; approval and disapproval have been expressed of that term.[16,41] Competitive and non-competitive inhibitions were established as occurring among specific pairs of substrates and products in a kinetic study of galactokinase[50] in which it was concluded that reaction had proceeded via such ternary complexes of the enzyme and its two substrates.

Inhibitory interactions were seen in the 1950s as playing major roles in the control of multi-enzyme systems. The factors first recognized as linking the stages of such systems were structural contiguity, appropriate K_m values, and inhibition of the enzyme catalysing an early stage by the product of a later stage.[41] New factors became evident with the discovery in 1955 of feedback inhibition in growing bacterial cells:[51] amino acids added to the cultures blocked the normal synthesis of a group of amino acids from other amino acids or from glucose as carbon source. Thus in the pathway of isoleucine synthesis from threonine, L-isoleucine inhibited the initial stage, threonine deaminase, by a specific and competitive process which, however, did not show straightforward Lineweaver–Burk relationships.[52]

In another such sequence, though inhibition appeared competitive, it was difficult to understand because of the great structural difference between the competing feedback inhibitor and the enzyme substrates. This applied to pyrimidine synthesis from aspartate, when a product, cytidine triphosphate, inhibited the initial stage, aspartate transcarbamylase.[53] Examined kinetically, an inhibition of 70% by 0·2 mM CTP with 5 mM aspartate was diminished to 5% when 25 mM aspartate was used; the maximum velocity was not affected. However, the inhibition of CTP proved labile and could be lost without diminution of transcarbamylase activity: treatment by heat or by urea had this action, showing the "feedback" site of CTP binding to be largely independent of the enzyme's catalytic sites. Intriguingly, ATP and CTP competed at the product-binding site; the evolutionary advantage offered by such sites was mooted. From the present point of view it is to be noted that structural similarity between competing substances had become firmly associated with competitive mechanisms (see Sections V and VI). Enzyme and substrate need not, however, fit in the lock and key fashion of Fischer, but may interact at active centres which are flexible and adopt an appropriate shape only during combination with potential substrates, as suggested in the induced-fit theory of Koshland.[16,54] Competitive and non-competitive relationships are still described as such and have been sub-classified according to the conformational changes involved.[54]

In further studies, competitive inhibitory relationships have been sub-divided into classical instances largely as expressed by Haldane,[6] caused by structural analogues and termed *isosteric*; and those caused by inhibitors with little structural resemblance to the substrate and termed *allosteric*.[25,54–56] The multiple binding sites often exhibited in allosteric systems can offer great flexibility in relationship between substrate and inhibitory or activating substances; the latter allow by co-operative effects, a change in the affinity of enzyme for substrate. It is noteworthy that haemoglobin gives an example of this phenomenon in its binding of oxygen, which was analysed many years ago. The possibility was raised, during these investigations,[55] of avoiding the "competitive" terminology by referring to K-system inhibition or V-system inhibition[55,57] to indicate the kinetic factors involved, but the use of "competitive" has remained; it is grounded in ideas of chemical structural relationships as well as of kinetics.

V. Applications in Microbiology

Following Haldane's[2,6] work, studies of competitive interaction were next applied in microbiology; they were designated as such, and their lineage from enzymology was overtly expressed. This resulted from an investigation of the action of sulphanilamide as an antibacterial agent by D. D. Woods[58] working in a Medical Research Council Unit which during the 1930s had identified a number of vitamin-like substances as nutrients essential for microbial growth. Woods was one of several investigators who reproduced *in vitro* the anti-bacterial actions on which therapeutic actions of the sulphonamides depended, by showing that they inhibited growth of bacterial cultures. He further found that such growth inhibition was antagonized by extracts of unknown composition from the micro-organisms being cultured, or from other plant or animal materials.

Using methods familiar in contemporary nutritional studies, Woods[58] fractionated the anti-sulphanilamide extract made by dilute ammonium hydroxide from several kg of yeast. From the evaporated extract, ethanol precipitated material inert as a sulphonamide antagonist and yielded a supernatant containing material which at 3 μg ml^{-1} of growth media, prevented sulphanilamide inhibition. The antagonist was further enriched by precipitation with mercuric acetate and removal of inert material with phosphotungstic acid. The active sulphonamide antagonist could be extracted from aqueous solutions of specified acidity to organic solvents, a property consistent with the antagonist having an acid pK of 4–5, as would a carboxylic acid. The concentrates were inactivated by nitrous acid and by acetylation, and their activity could be regenerated from the acetylated material by alkaline hydrolysis. These properties were consistent with the presence of an amino group, but the antagonist was soluble in ether and thus the amino group did not appear to belong to an aliphatic α-amino acid. Suspecting an aromatic amino acid, the product from nitrous acid was caused to react with dimethyl α-naphthylamine, and then gave a red colour suggesting diazotization and coupling. It was concluded that, although other possibilities remained, the antagonist could be an aromatic amine and carboxylic acid, so presenting structural similarity to sulphanilamide itself.

Examination of the properties of the antagonist in tests of microbial growth had, contemporaneously, given the results of Fig. 3a. Thus[58]

> . . . it was found that if the concentration of sulphanilamide was increased, it is necessary to raise the concentration of the (antagonist) in the same proportion in order to reverse the inhibition . . . this constant quantitative relationship between inhibitor and active substance was reminiscent of the competitive inhibition of enzyme reactions by substances chemically related to the substrate or product.

This was compared with the inhibition of invertase by stereochemically similar monosaccharides, of lipases by carbonyl group compounds, and of succinic dehydrogenase by malonates, with Haldane's[6] monograph as the only reference cited in this connection. Haldane was indeed the only author who had brought together these diverse studies (see Section IV) as examples of competitive inhibition. This additional evidence for the antagonist being a structural analogue of sulphanilamide led to its attempted replacement by p-aminobenzoic acid. The compound was indeed successful in minute amounts in antagonizing sulphanilamide.

This admirable conjunction of chemical and enzymic knowledge had demonstrated a novel fashion in which principles of competitive inhibition could be applied: in constructing hypotheses on the existence and structure of naturally-occurring metabolites corresponding to established therapeutic agents. A further application was promptly made:[59] starting with known microbial growth essentials, structural analogues were examined as antimicrobial agents and a number found to be inhibitory. Thus pyridine-3-sulphonic acid amide (PSA) bore the same relationship to nicotinic acid that sulphanilamide did to p-aminobenzoic acid; PSA and the sulphonic

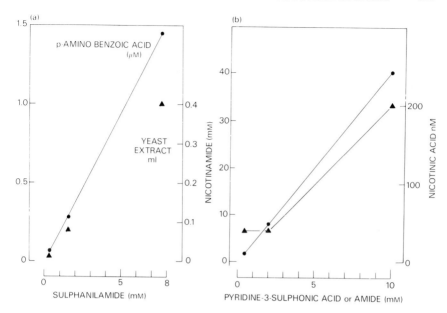

Fig. 3. Competitive relationships on bacterial growth. (a) A strain of *Streptococcus haemolyticus* was grown in a medium of known substances but its growth inhibited in some tubes by the concentrations of sulphanilamide shown as abscissae. The ordinates give the concentrations of added substances found necessary to allow growth to proceed in the presence of the sulphanilamide (ref. 58). (b) A staphylococcus strain was grown in defined media which contained nicotinic acid, ▲, or its amide, ●, and in addition the concentrations of pyridine sulphonic acid, ▲, or its amide, ●, shown on the abscissae. The ordinates give the minimal concentrations of nicotinic acid or its amide found necessary for growth (ref. 59).

acid were found to inhibit the growth of several micro-organisms. Inhibition by the amide showed competitive relationships with nicotinamide; relationships between the acids were partly competitive (Fig. 3b). Many further examples followed, some showing chemotherapeutic action *in vivo*,[60] and others invoking a wide range of structural analogues.[19,61] Generalization from such data gave insight into how, in evolution, structural variants which arose by mutation could by their properties as enzyme inhibitors be of offensive or defensive value to the organisms producing them.[62]

In most cases quoted the quantitative data supporting the description of competitive relationships were given by relatively simple growth tests. Subsequently, Woods and colleagues[63] demonstrated, by measuring the folic acid formed by *Streptobacterium plantarium* from *p*-aminobenzoic acid in the presence of varying concentrations of sulphonamides, that "the reversal of p-aminobenzoic acid was of a strictly competitive nature". This data may not be adequate to meet criteria developed later (Section IV.C); concepts from enzyme competition have, however, remained in active use in chemotherapeutic work. These applications of competitive relationships involved responses of whole organisms to chemical agents: a category of test system not involved in Haldane's[6] monograph, but included in his 1927 paper[2] (Section II.A). The competitive nomenclature thus established itself in biology with the use of representations such as those of Figs 1, 2 and 3 rather than with Lineweaver–Burk plots and their successors.

VI. Applications in Toxicology and Pharmacology

Claude Bernard's[64] study in 1856 of the interactions of O_2 and CO with the haemoglobin of mammalian blood featured prominently in his lectures on medicaments and toxic agents; it occupied four of the book's 30 lectures and was given in delightful experimental detail. His conclusions are expressed in terms of the *displacement* of O_2 by CO, which was measured in mm^3 by techniques of gas analysis. For Bernard was unaware of the reverse relationship: CO ". . . instantly displaces the oxygen of the red corpuscles and cannot itself be subsequently displaced by oxygen". Thus Bernard's study does not rank as a recognition of competitive relationships, which in relation to CO and O_2 were established by J. S. Haldane and colleagues (Section II.B). Bernard's work ensured, however, that this action of CO was well known scientifically and his term "displacement" has been used by Warburg (Section II.A), by J. B. S. Haldane (Section II.B) and subsequently in situations when "competitive" would also have been applicable. Moreover, it contributed to the view that substances active as drugs combined with components of the living systems affected by them, as expressed in Ehrlich's aphorism *"corpora non agunt fixita"*: substances do not act unless bound.[65]

A. DRUG ANTAGONISM

A terminology of "antagonism" was used initially to describe interactions of drugs or humoral agents when one depressed the action of another. Such was to be found in the interaction of adrenaline and ergotamine reported by Broom & Clark.[66] The transition from this to more specific use of "competitive" terminology is to be found particularly in the work of A. J. Clark, who explored mathematical relationships between degree of drug action, concentration and time factors in a wide range of toxic and pharmacological agents in his enterprising monograph of 1933 entitled *The Mode of Action of Drugs on Cells.*[67]

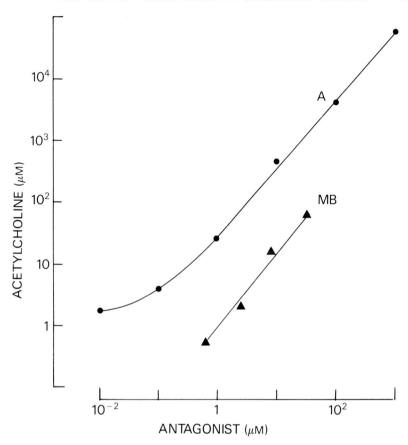

Fig. 4. The concentrations of atropine, A, or of methylene blue, MB, which gave 50% diminution of the response of frog heart, caused by the concentrations of acetylcholine given as ordinates (refs 68, 69).

In his earlier work on this theme, Clark[68] described as "antagonism" the relationship between acetylcholine and atropine in the experiments illustrated in Fig. 4. With preparations of frog rectus abdominis and heart, a given ratio of concentrations of acetylcholine to atropine over a 1000-fold range gave a nearly constant degree of action. Clark[68] noted that this relationship was similar to that between CO and O_2 at haemoglobin as recorded by Douglas et al.,[7] and like those authors he did not use "competitive" terminology.

Shortly afterwards, at Clark's suggestion and with his help, Cook[69] measured interactions of acetylcholine and methylene blue, including those of Fig. 4. This showed that, above threshold values, [acetylcholine]/

[methylene blue][n] for equal actions was a constant, and that over a 100-fold range of concentrations, n did not differ significantly from unity.

Methylene blue coloured the tissue, but this was irreversible and unrelated quantitatively to the concomitant pharmacological action. Acetylcholine action was by contrast reversible, as was that of methylene blue on tissue contraction. This, therefore, concerned "some freely reversible reaction". Again, the agents were described as antagonists. By Haldane's[6] criteria they would have been described as competing: Haldane's book[6] concerned enzymes and Cook's[69] work was not included.

A subsequent Discussion Meeting[70] on the chemical and physical basis of pharmacological action in which several such findings were quoted, emphasizes the time-lag in the use of competitive nomenclature. The eight investigators whose contributions are reported included A. J. Clark, W. Straub, R. A. Peters, J. H. Quastel, H. R. Ing and J. H. Gaddum, some of whom later made much use of "competitive" terms. The report does not, however, include the use of the terms, though examination of quantitative drug interactions and of common ground between biochemistry and pharmacology is advocated, and interpretation of concentration/action curves by mass-action law (Section III) is noted.

In the meantime, a significantly limited use of "competitive" terminology in pharmacology had been made. A. J. Clark's[67] monograph dealt with many aspects of the action of drugs on cells, including theories derived from equilibria and the kinetics of binding. However, in describing antagonism among drugs, competitive terminology was employed only in describing the work of Haldane and of Warburg on the "antagonism between oxygen and carbon monoxide. The two gases both form reversible compounds with haemochromogens and compete for the same receptors". Cyanides were described as combining with a receptor different from that with which O_2 combined, but the term non-competitive was not employed.

B. SULPHONAMIDES AND COMPETITION

This tentative introduction of "compete" to pharmacological writing was not maintained in the Discussion Meeting[70] nor in a textbook of 1940: J. H. Gaddum's[71] *Pharmacology* which quoted Clark's writings, described drug antagonisms without specifying competition, although the carbon monoxide–haemoglobin interaction was included, and also sulphanilamide as an anti-bacterial agent. In a subsequent edition, however,[71] the account of sulphonamides is placed at the beginning of a section on chemotherapy, and Wood's[58] study of their action is included with the description "this is an example of substrate competition". Moreover, work similar to that cited above is now described differently: "the adrenaline and ergotamine are probably competing with one another for receptors . . ."; and "the same kind of competition probably occurs between eserine and acetylcholine for the enzyme cholinesterase, and between atropine and acetylcholine for muscarine receptors . . .".

The immediate source of Gaddum's use[71] in 1953 of competitive nomenclature thus appears to be Wood's[58] account of p-aminobenzoate. It does not appear likely that he regarded "competition" as being confined to enzyme studies, in view of his applying it to muscarinic receptors. Significantly, Gaddum[71] did not adopt Clark's[67] use of "competition" in relation to CO/O_2: possibly because Clark[67] placed this interaction in one category, and that of ergotamine/adrenaline in another. Gaddum[72] had indeed earlier offered a different explanation for the ergotamine relationship in terms of tissue heterogeneity; thus his later description[71] of the relationship as competitive reflected the explanatory power seen to be exerted by the then considerable corpus of "competitive" theory and examples. This value was seen also by others. Dates of publication of cognate works featuring competitive nomenclature are: J. H. Burn[73] on sulphonamides, 1948; Work & Work[74] and Findlay[75] on chemotherapeutic agents, 1948 and 1950; Barlow & Ing[76] regarding acetylcholine with tetraethylammonium salts and histamine with antihistaminics, 1955. Thus all can derive from Haldane[6] through Woods.[58]

C. BINDING STUDIES, RECEPTORS AND COMPETITION

In his pioneering studies on neurotransmission, J. N. Langley[77] has been credited[78] with concluding that nicotine and curare "both must compete at the same 'receptive substance' or receptor". This is not, however, the terminology of Langley's 1906 Croonian lecture[77] which is cited and in which nicotine is described as paralysing motor nerve endings, while there is mutual *antagonism* with curare.

> The mutual antagonism of nicotine and curare on muscle can only satisfactorily be explained by supposing that both combine with the same radicle of the muscle, so that nicotine–muscle compounds and curare–muscle compounds are formed. Which compound is formed depends on the mass of each poison present and the relative affinities for the muscle radicle.

Langley is concerned, rather, with "nerve endings and special excitable substances in cells" and with another distinction:[77]

> the muscle substance which combines with nicotine or curare is not identical with the substance that contracts . . . receptive substance receives the stimulus and by transmitting it, causes contraction . . . the substances, I take it, are radicles of the protoplasmic molecule.

Langley[77] thus proposed an early application of mass-action principles (Section III) to a biological system, but did not give an analysis comparable to that of Section II which would be needed to justify the subsequent application of the term "compete". Langley was consistent, in his subsequent writing, in the use of the terms antagonism, prevention, stops or reduces, but not of competition, to describe pharmacological interactions which included the nicotine–curare instance.[79,80] Entry of the competitive terminology to these subjects thus remains as described in Section VI.B above. Laduron[78] is, however, one of several workers who apply competitive terminology in receptor binding studies and indeed offers as one definition of a receptor "a site of competition for agonist and antagonist (when) the stimulus produced by the agonist through a mechanism not yet elucidated leads to a physiological response". The receptor site is thus more specific than acceptor sites which do not have a necessary connection with such response. Measurement of both categories of sites has become much easier with the development of potential ligands which are highly labelled isotopically.

The Scatchard analysis[81,82] frequently used in such work derives from the same basic considerations as does the Michaelis–Menten equation (Section II.D), but depends on measurement of the amount B of a ligand bound (total binding less non-specific) at a concentration F of free ligand. The Scatchard plot is of B/F versus B, and can yield a maximum binding B_{max} and the dissociation value K_d of the ligand. Recently, interactions of D-2 dopamine receptor preparations from rat striatum were so analysed by Hamblin et al.[83] Competition was described as resulting from interactions with metoclopramide and other agents; terminology of "displacement" was also used. The interactions occurred in a fashion best explained by the D-2 receptor existing in two interconverting states of differing K_d; the competition was affected by Na^+, by temperature and by guanine nucleotides,[83] observations reminiscent of cyclic AMP studies (see also Section IV.C).

Allosteric considerations similar to those of Section IV.C are invoked also in relation to examples of receptor inhibition, when these show variable degrees of competitive relationships. Cerebral cortical membrane preparations were shown by Scatchard plots to be labelled by a convulsant $[^{35}S]$ phosphorothionate in a partly competitive, multi-component fashion; barbiturates accelerated the dissociation of the ^{35}S compound from its binding sites which were concluded to be allosterically linked to sites binding the barbiturate.[84]

Competitive binding techniques and terminology have been applied also in investigating sensory receptors: a $[^3H]$-labelled methoxypyrazine smelling powerfully of pepper was found to bind specifically and saturably to a membrane preparation from sensory areas of cow nasal epithelium.[85] In a system of membrane protein with the pyrazine and variable quantities of other pepper-smelling compounds, these latter inhibited binding competitively and at concentrations which were correlated with the detection thresholds for the odorants.

VII. Competitive Terminology: Technical and Non-technical

The competitive terminology for interactions among chemical substances at biological systems thus remains in active use some 60 years after its initiation; it has been applied in new types of investigation and has been subject to subclassification but not replacement. It retains as basic concepts the spatial distribution of atoms and affinities in substrates, inhibitors and enzymes or binding-proteins to which Fischer, Armstrong, J. S. Haldane, Ehrlich, Langley, Michaelis and J. B. S. Haldane contributed. In particular, the linkage of competitive and non-competitive to the enzyme kinetics of Michaelis and of his successors gave those terms a technical exactitude and firmness in application, which contrasted with the use of "antagonism", which retained a more general use. This more technical use of "competitive" remains even though the exact application of enzyme kinetics or cognate expressions is only occasionally invoked. The connection with mathematical expressions for enzyme kinetics or binding also made the use of "competitive" in biochemistry and pharmacology more exacting than when used in biology, or generally. It is, however, from the manner and date of this more general use that the adoption of competitive terminology in biochemistry can be understood.

The idea of competition is inherent in Darwin's concept of evolution and he used the term accordingly:[86] "individuals of the same species come in all respects into the closest competition with each other"; "the slightest advantage in one . . . over those with which it comes into competition . . . will turn the balance". Competition has remained an active concept in ecological studies; to quote an example contemporary with Warburg and Haldane's enzyme studies, Tansley[87] writes of competition between plants, and competition as affected by animal attack: ". . . only those species can be present which are able to exist under the given conditions and in competition with the other species present".

Darwin and darwinists owed an important part of their ideas to writers on social and economic subjects. Adam Smith in 1776 noted economic progress when ". . . regulation is by competition and the market place and not by the state".[88,89] In 1798 Malthus,[90] who was cited by Darwin,[86] made much use of competitive terminology in discussing political economy. Free competition is contrasted with monopoly in its effect on price: "some sacrifice must be made by the competitors"; and on wages: "abundance and competition always have a relative significance".[91] Following the successful development of cognate ideas by Darwin, doctrines of "social darwinism" developed in the decades around 1880 (see Galbraith[89]) in which competitive ideas and terminology dominated. T. H. Huxley[92] wrote of industrial competition and of "this eternal competition of man against man and of nation against nation"; "struggling for existence with competitors, every ignorant person tends to become a burden upon and an infringer of the liberty of his fellows and an obstacle to their success". Competition in careers and competitive examinations which had been in operation in certain UK professions since the 1850s, were noted by Galton.[93] It is not surprising that a behaviourist (Skinner[94]) wrote that science emerged from a competitive society.

Current writers, however, report that despite the prolific Victorian writing on competition, it was in subsequent periods (closer to its introduction to biochemical use) that discussion intensified. Thus debates on competition and price policies are noted by Marshall[95] as prompted by the trade depressions and the unemployment of the 1920s and early 1930s. Galbraith[89] also observes: "In the closing decades of the last century and the early years of the twentieth, economists became increasingly preoccupied with the operation of the model of a competitive society"; and "in the thirties there was an especially widespread effort to mitigate undesirable economic effects of insecurity associated with competitive market prices" and the uncertainties of market competition.

Economic theory has also included concepts of competition closer to the biochemical use of the term. Thus equilibrium in a market is described as favoured by stability of product, price, and number of competing firms.[95] The competition has been described as perfect or imperfect, situations which show some resemblance to competitive and non-competitive enzyme kinetics. Among factors potentially increasing the degree of competition is the availability of competitive substitutes for established products, for example articles in synthetic materials replacing those in metal or leather. These can affect the point of equilibrium among market competitors and thus present in part a parallel to the competitive substrate-analogues of enzymology.

Moreover, the 1920s and 1930s saw a marked increase in popular newspaper competitions, and in the competitive organization of sports. Specific reference is made to competition at this period in entertainment and between newspapers, in Allen's[96] history of a British newspaper. Such is the background to the inhibitory categories which Michaelis and co-workers recognized in 1913, being given the apt and expressive competitive terminology only later, in 1927 and 1930. J. B. S. Haldane, who conferred this gift, was shortly afterwards writing also[97] on economic matters, sociology and ecology and employing the term competition in its non-biochemical senses.

Also in the 1930s, processes of scientific discovery were described[98,99] as proceeding by the proposing and examining of theories to find those which "stood up to tests better than their competitors", the appropriate being selected in near-darwinian fashion.

It may finally be queried whether there is a common functional role for processes described as competitive in the many spheres of knowledge and activity in which the term has been applied. The preceding accounts suggest that there is, and that the common role concerns the participation in relatively autonomous regulatory processes which produce from a multiplicity of simpler competing units, a relatively stable and more complex organization: the animal and plant communities of ecology; in human social and economic activities; in thought and sensory processes as well as in metabolic control. The entities which compete in feedback processes of metabolic control can be seen as becoming integrated to more complex multi-component systems, with the increased sensitivity and responsiveness which such organization confers.

VIII. Summary

(1) The terms competition and competitive were in use for appropriate types of interaction in human and animal behaviour from the seventeenth century. In the nineteenth and early twentieth centuries they reached more technical uses in biology, especially in darwinian studies; and in chemistry in describing competing reactions, surface phenomena and the influence of substituent groupings in reactant molecules.

(2) Use of competitive and non-competitive to describe enzyme inhibitors had a specific beginning when J. B. S. Haldane[2,6] (following premonitory work of others[3,23]) applied the terms in 1927 and 1930 to types of inhibition already differentiated by Michaelis and co-workers.[20–22] The theoretical background in kinetics and stereochemistry so acquired gave a firmness to the application of the terms in biochemistry. The first examples concerned glycosidases, especially β-D-fructofuranosidase or invertase, and interactions of carbon monoxide and oxygen at iron–porphyrin systems. They were thus of interest in toxicology and in enzyme and carrier studies.

(3) The sphere of application of the biochemically-defined terms expanded greatly when, following investigation of sulphonamide action, it was realized that concepts of enzyme inhibition by structurally related compounds offered a route to understanding the action of existing medicaments and to the production of new ones.

(4) Ideas and terminology based on competitive and non-competitive enzyme inhibition and receptor occupancy have subsequently been applied

in many ways. Examples include application to the analysis of feedback inhibition and other processes of metabolic control; to receptor relationships among neurotransmitters and medicaments; and to understanding interactions at sensory receptors.

ACKNOWLEDGEMENTS

In preparing this essay I have been greatly indebted to Professor H. S. Bachelard for facilities in this Department, to the Wellcome Trust for support, and to G. B. Ansell, H. S. Bachelard, M. A. McIlwain, J. H. Thomas and M. B. Thorn for their helpful comments.

REFERENCES

1. The Compact Edition of the Oxford English Dictionary (1971), p. 490. Clarendon, Oxford.
2. Haldane, J. B. S. (1927). Carbon monoxide as a tissue poison. *Biochem. J.* **21**, 1068–1072.
3. Warburg, O. (1926). Uber die Wirkung des Kohlenoxyd auf den Stoffwechsel der Hefe. *Biochem. Z.* **177**, 471–486.
4. Schoenewald, F. S. (1949). *German–English Medical Dictionary*. Lewis, London.
5. Warburg, O. (1949). *Schwermetalle als Wirkungsgruppen von Fermenten*, pp. 70–81. Cantor, Freiburg.
6. Haldane, J. B. S. (1930). *Enzymes*. Longmans, Green, London. (Reprint: MIT Press, Cambridge, MA, 1965.)
7. Douglas, C. G., Haldane, J. S. & Haldane, J. B. S. (1912). The laws of combination of haemoglobin with carbon monoxide and oxygen. *J. Physiol.* **44**, 275–304.
8. Haldane, J. & Lorrain Smith, J. (1896). The oxygen tension of arterial blood. *J. Physiol.* **20**, 497–520.
9. Foster, C. le N. & Haldane, J. S. (ed.) (1905). *The Investigation of Mine Air*. Griffin, London.
10. Haldane, J. S. (1917). *Organism and Environment as Illustrated by the Physiology of Breathing*. Yale University Press, New Haven.
11. Haldane, J. S. (1922). *Respiration*, pp. 72–83; 320–332. Yale University Press, New Haven.
12. Haldane, J. S. (1895). The relationship of the action of CO to O_2 tension. *J. Physiol.* **18**, 201.
13. O'Sullivan, C. & Tomson, F. W. (1890). Invertase: a contribution to the history of an enzyme or unorganized ferment. *Chem. Soc. Trans.* **57**, 834–931.
14. Duclaux, E. (1899). *Traite de Microbiologie*, Tome II: *Diastases, Toxines et Venins*. Masson, Paris.
15. Kuhne, W. (1878). Erfahrungen und Bemerkungen über Enzyme und Fermente. *Unters. aus dem physiol. Institut der Univer. Heidelberg* **1**, 291–326.
16. Cornish-Bowden, A. (1979). *Fundamentals of Enzyme Kinetics*, pp. 16–37; 73–118; 147–176. Butterworths, London.

17. Fischer, E. (1894). Einfluss der Konfiguration auf die Werkung der Enzyme. *Ber. deut. chem. Ges.* **27**, 2985–2995.
18. Fischer, E. & Armstrong, E. F. (1901). Über die isomeren Acetohalogen-Derivate des Traubenzuckers und die Synthese der Glucoside. *Ber. deut. chem. Ges.* **34**, 2885–2903.
19. Woolley, D. W. (1952). *A Study of Antimetabolites.* Wiley, New York.
20. Michaelis, L. & Pechstein, H. (1914). Über die verschiedenartige Natur der Hemmungen der Invertasewirkung. *Biochem. Z.* **60**, 79–90.
21. Michaelis, L. & Menten, M. L. (1913). Die Kinetik der Invertinwirkung. *Biochem. Z.* **49**, 333–369.
22. Michaelis, L. & Rona, P. (1914). Die Wirkungsbedingungen der Maltase aus Bierhefe. III. Über die Natur der verschiedenartige Hemmungen der Fermente. *Biochem. Z.* **60**, 62–78.
23. Armstrong, E. F. (1919). *The Simple Carbohydrates and the Glucosides.* Longmans, Green, London.
24. Nelson, J. M. & Anderson, R. (1926). Glucose and fructose retardation of invertase action. *J. Biol. Chem.* **69**, 443–448.
25. Dixon, M. & Webb, E. C. (1979). *Enzymes,* 3rd edn. Longmans, Green, London.
26. Pattison Muir, M. M. (1884). *A Treatise on the Principles of Chemistry,* pp. 369–443. University Press, Cambridge.
27. Robinson, R. (1932). *Outline of an Electrochemical (electronic) Theory of the Course of Organic Reactions,* pp. 39–46. Institute of Chemistry, London.
28. Remick, A. E. (1943). *Electronic Interpretations of Organic Chemistry,* pp. 95–102. Wiley, New York.
29. Adam, N. K. (1938). *The Physics and Chemistry of Surfaces,* pp. 58–101. Clarendon, Oxford.
30. Smith, J. M. (1965). Obituary, J. B. S. Haldane. *Nature (London)* **206**, 239–240.
31. Briggs, G. E. & Haldane, J. B. S. (1925). A note on the kinetics of enzyme action. *Biochem. J.* **19**, 338–339.
32. Murray, D. R. P. (1929). Molecular constitution and accessibility to enzymes. The effect of various substances on the velocity of hydrolysis by pancreatic lipase. *Biochem. J.* **23**, 292–308.
33. Velluz, M. L. (1927). Recherches sur l'action inhibitrice. *Bull. Soc. Chim. Biol.* **9**, 483–500.
34. Grassman, W. & Klenk, L. (1930). Zur Frage der Einheitlichkeit tiersche und pflanzlicher Dipeptidase. *Z. Physiol. Chem.* **186**, 26–49.
35. Harden, A. & Zilva, S. S. (1915). The reducing enzyme of *B. coli communis. Biochem. J.* **9**, 379.
36. Stephenson, M. (1930). *Bacterial Metabolism,* pp. 36–39. Longmans, Green, London.
37. Quastel, J. H. & Wooldridge, W. R. (1928). Some properties of dehydrogenating enzymes of bacteria. *Biochem. J.* **22**, 689–702.
38. Quastel, J. H. & Wooldridge, W. R. (1927). Experiments in bacteria in relation to the mechanism of enzyme action. *Biochem. J.* **21**, 1224–1251.
39. Webb, J. L. (1966). *Enzyme and Metabolic Inhibitors,* Vol. 2, p. 259. Academic Press, New York.
40. Cook, R. P. (1930). A comparison of the dehydrogenations produced by *B. coli communis* in the presence of oxygen and methylene blue. *Biochem. J.* **24**, 1538–1550.

41. Dixon, M. & Webb, E. C. (1958). *Enzymes*. Longmans, Green, London.
42. Ansell, G. B. (1983,4). Judah H. Quastel at 84. *Biochem. Soc. Bull.* **5**, 3–4; and **6** (2), 11.
43. Hudson, A. J. (1984). The first Rossiter Research Conference. *Int. Brain. Res. Organization News* **12**, 13.
44. Quastel, J. H. (1981). A brief autobiography. *Bull. Canad. Biochem. Soc.* **18**, 13–34.
45. Woolf, B. (1932). In *Allgemeine Chemie der Enzyme* (Haldane, J. B. S. & Stern, K. G., eds), p. 119. Steinkopff, Dresden.
46. Lineweaver, H. & Burk, D. (1934). The determination of enzyme dissociation constants. *J. Am. Chem. Soc.* **56**, 657–666.
47. Thorn, M. B. (1961). Enzyme kinetics. In *Biochemists' Handbook* (Long, C., ed.), pp. 205–218. Spon, London.
48. Ebersole, E. R., Guttentag, C. & Wilson, P. W. (1943). The nature of carbon monoxide inhibition of biological nitrogen fixation. *Arch. Biochem.* **3**, 399.
49. Dodgson, K. S., Spencer, B. & Williams, K. (1956). Examples of anti-competitive inhibition. *Nature* (*London*) **177**, 432–433.
50. Gulbinsky, J. S. & Cleland, W. W. (1968). Kinetic study of *E. coli* galactokinase. *Biochemistry* **7**, 566–575.
51. Newsholme, E. A. & Start, C. (1973). *Regulation in Metabolism*. Wiley, London.
52. Umbarger, H. E. (1956). Evidence for a negative feedback mechanism in the biosynthesis of isoleucine. *Science* **123**, 848–9.
53. Gerhardt, J. C. & Pardee, A. B. (1962). The enzymology of control by feedback inhibition. *J. Biol. Chem.* **237**, 891–896.
54. Koshland, D. E. (1970). Molecular basis for enzyme regulation. In *The Enzymes*, 3rd edn (Boyer, P. D., ed.), pp. 341–396. Academic Press, New York.
55. Monod, J., Wyman, J. & Changeaux, J.-P. (1965). On the nature of allosteric transitions: a plausible model. *J. Mol. Biol.* **12**, 88–118.
56. Slater, E. C. & Bonner, W. D. jun. (1952). The effect of fluoride on the succinic oxidase system. *Biochem. J.* **52**, 185–196.
57. Stadtman, E. R. (1970). Mechanisms of enzyme regulation in metabolism. In *The Enzymes*, Vol. 1, (Boyer, P., ed.), pp. 397–459. Academic Press, New York.
58. Woods, D. D. (1940). The relation of *p*-aminobenzoic acid to the mechanism of the action of sulphanilamide. *Br. J. Exp. Pathol.* **21**, 74–89.
59. McIlwain, H. (1940). Pyridine-3-sulphonic acid and its amide as inhibitors of bacterial growth. *Br. J. Exp. Pathol.* **21**, 136–147.
60. McIlwain, H. & Hawking, F. (1943). Chemotherapy by blocking bacterial nutrients. *Lancet* **i**, 449–451.
61. McIlwain, H. (1943). The anti-streptococcal action of iodinin. Naphthaquinones and anthraquinones as its main natural antagonists. *Biochem. J.* **37**, 265–271.
62. McIlwain, H. (1944). Origin and action of drugs. *Nature* (*London*) **153**, 300–304.
63. Nimmo-Smith, R. H., Lascelles, J. & Woods, D. D. (1948). Synthesis of "folic acid" by *Streptobacterium plantarum* and its inhibition by sulphonamides. *Br. J. Exp. Pathol.* **29**, 264–279.
64. Bernard, C. (1856). *Leçons sur les effets des substances toxiques et medicamenteuses*. Bailliere, Paris.
65. Farmer, S. G. (1985). *Corpora non agunt fixita. Trends Pharmacol. Sci.* **6**, 99–100.

66. Broom, W. A. & Clark, A. J. (1923). *J. Pharmacol. Exp. Ther.* **22**, 59.
67. Clark, A. J. (1933). *The Mode of Action of Drugs on Cells*, pp. 228–240. Arnold, London.
68. Clark, A. J. (1926). The antagonism of acetylcholine by atropine. *J. Physiol.* **61**, 547–556.
69. Cook, R. P. (1926). The antagonism of acetylcholine by methylene blue. *J. Physiol.* **62**, 160–165.
70. Discussion Meeting. (1937). On the chemical and physical basis of pharmacological action. *Proc. R. Soc. B.* **121**, 580–609.
71. Gaddum, J. H. (1940, 1953). *Pharmacology*. Oxford University Press, London.
72. Gaddum, J. H. (1926). The action of adrenalin and ergotamine on the uterus of the rabbit. *J. Physiol.* **61**, 141–150.
73. Burn, J. H. (1948). *Background to Therapeutics*. Oxford University Press, London.
74. Work, T. S. & Work, E. (1948). *The Basis of Chemotherapy*. Oliver and Boyd, Edinburgh.
75. Findlay, G. M. (1950). *Recent Advances in Chemotherapy*, 3rd edn. Churchill, London.
76. Barlow, B. B. & Ing, H. R. (1955). *Introduction to Chemical Pharmacology*. Methuen, London.
77. Langley, J. N. (1906). Croonian Lecture. On nerve endings and special excitable substances in cells. *Proc. R. Soc. B* **78**, 170–194.
78. Laduron, P. M. (1984). Criteria for receptor sites in binding studies. *Biochem. Pharmacol.* **33**, 833–839.
79. Langley, J. N. (1907). On the contraction of muscle. *J. Physiol.* **36**, 347–384.
80. Langley, J. N. (1921). *The Autonomic Nervous System*, pp. 1–80. Heffer, Cambridge.
81. Scatchard, G. (1949). The attractions of proteins for small molecules and ions. *Ann. N.Y. Acad. Sci.* **51**, 660–672.
82. Yamamura, H. I., Enna, S. J. & Kuhar, M. J. (1978). *Neurotransmitter Receptor Binding*. Raven, New York.
83. Hamblin, M. W., Leff, S. E. & Creese, I. (1984). Interactions of agonists with D-2 dopamine receptors. *Biochem. Pharmacol.* **33**, 877–887.
84. Trifiletti, R. R., Snowman, A. M. & Snyder, S. H. (1984). Barbiturate recognition site. *Europ. J. Pharmacol.* **106**, 441–447.
85. Pevsner, J., Trifiletti, R. R., Strittmatter, S. M. & Snyder, S. H. (1985). Isolation and characterization of an olfactory receptor protein for odorant pyrazines. *Proc. Natl. Acad. Sci. USA* **82**, 3050–3054.
86. Darwin, C. (1859). *On the Origin of Species by means of Natural Selection, or the Preservation of Favoured Races in the Struggle for Life*, pp. 420–421. University Press, Oxford (1902 reprinting).
87. Tansley, A. G. (1923). *Practical Plant Ecology*. London: Allen & Unwin (pp. 19, 73 and 77 of 1926 reprinting).
88. Smith, A. (1776). *An Inquiry into the Nature and Causes of the Wealth of Nations*, Ch. 8.
89. Galbraith, J. K. (1977). *The Affluent Society*, pp. 18–28. Deutsch, London.
90. Malthus, T. R. (1798). *An Essay on the Principle of Population as it Affects the Future Improvement of Society*. Johnson, London.
91. Malthus, T. R. (1824). Essay on Political Economy. *Q. Rev.* **30**, 297–334; reprinted in (1963) *Occasional Papers of T. R. Malthus* (Semmel, B., ed.), pp. 171–208. Franklin, New York.

92. Huxley, T. H. (1888). *The Struggle for Existence in Human Society*; reprinted in (1906) *Evolution and Ethics* (Huxley, T. H.), pp. 195–235. Macmillan, London.
93. Galton, F. (1883, 1907). *Inquiries into Human Faculty and its Development.* Dent, London (pp. 200–220 of 2nd edn, 1907).
94. Skinner, B. F. (1948). *Walden Two.* Macmillan, New York.
95. Marshall, B. V. (1975). *Comprehensive Economics, Institutional Analytical and Applied*, pp. 903–954; 963–972; 999–1008. Longman, London and New York.
96. Allen, R. (1983). *The Voice of Britain.* Stephens, Cambridge.
97. Haldane, J. B. S. (1938). *The Marxist Philosophy and the Sciences*, pp. 93–127; 154–176. Allen & Unwin, London.
98. Popper, K. (1934). *Logik der Forschung.* Julius Springer, Vienna.
99. Popper, K. (1976). *Unended Quest: an Intellectual Autobiography*, pp. 86–87; 103–104; 148–151. Collins, Glasgow.

Subject Index

A

N-Acetyllactosamine synthetase
  epididymal, 19, 20
  in biosynthesis of glycoproteins, 4
  in sperm maturation, 19
  interaction
    with α-lactalbumin, 4, 5, 16, 19, 20, 22
    with α-lactalbumin-like protein, 19, 22
  involvement in sperm-egg binding, 20
  localization in Golgi, 4
  size, 5
Acetylcholine, 73, 74, 79, 84, 93, 94, 176, 177
  competition
    from atropine, 174–175
    from methylene blue, 174–175
Acetylcholinesterase, 93, 94, 103, 104
Actinomyces junthinus protease
    inhibitor, fluorescence maximum, 139, 140
Activation energy, dependence on
    dipolar interactions, 148–150
Adrenaline, 72, 174, 176, 177
Adrenal medulla, enkephalins, 72, 74, 76, 78, 87, 88, 89
Albumin, fluorescence maximum, 139, 140
Alcohol dehydrogenase, 127
Aldolase, 127
  red-edge effect, 142
Alkaline phosphatase, 127
Alzheimer's disease, 110
γ-Aminobutyrate, 70, 74, 84, 93, 94
Aminopeptidase
  brain, 101
  microvillar,
    A, properties, 96
    N, properties, 96
    W, properties, 96
Angiotensin, 71, 103, 106, 110
Angiotensin converting enzyme (ACE)
    see peptidyl dipeptidase A

Ankylosing spondylitis, 55
Antithrombin III, 49
α-Antitrypsin inhibitor, in
    proalbuminaemia, 86–87
A^1-protein, of lactose synthase, 2, 22
Aspartate transcarbamylase, inhibition
    of, 170
Atrial natriuretic factors (ANF), 71, 72, 83
Atriopeptins *see* atrial natriuretic
    factors
Atropine, 174, 175
Azurin, 127, 128, 139

B

Biogenic monoamines, 70, 73
Bombesin, 71
B-protein, of lactose synthase, 2, 22
Bradykinin, 71, 99, 103, 106, 109
Breast carcinomas, response to
    endocrine therapy, 16

C

C1 complex, of complement system
  activation mechanism, 34–36
  control by C1-inhibitor, 36, 47–48
  electron microscopy, 31, 33, 34
  model of, 34
  neutron diffraction studies, 33, 34
  structure of, 31, 32
  subcomponents
    amino acid sequence, 31
    interaction of, 33–36
    properties of, 33, 35
    structure of, 31–33
C3 convertase(s)
  decay of, 42
  effect of,
    control proteins, 42

C3 convertase(s) *continued*
  membrane receptors, 42
  formation of, 36–39, 41–42, 48
C5 convertase(s)
  formation of, 43–44
Calcitonin, 71, 73, 82–83
  gene, differential expression, 82–83
Calcitonin gene-related peptide
    (CGRP), 71, 73, 82–83
Calmodulin, 6
Carbonic anhydrase, red-edge effect,
    142
Carboxyhaemoglobin, 160, 161, 162
Carboxypeptidase A, 104, 105–106
  catalytic mechanism, 105
  inhibitors of, 105
  properties of, 105
  synthesis of, 105
Carboxypeptidase B (protaminase;
    EC 3.4.17.2), 86, 88
Carboxypeptidase E (enkephalin
    convertase),
  assay for, 88
  localization of, 88–89
  properties of, 88–89
  purification of, 88
Casein(s), 1, 4, 7, 9, 10, 11, 12, 18
Catecholamines, 84, 91, 94, 95
Catechol-*O*-methyl transferase, 94
Cathepsin B, 87
Chemotactic peptide, 99
Cholecystokinin, 71, 72, 73, 74, 88, 91,
    99, 101, 107, 110
  proteolytic processing, 88
Chymotrypsin, red-edge effect, 142
Competition
  antonyms to, 159
  between O_2 and CO
    in human respiration, 159–161
    in yeast respiration, 159–161
  definition of, 159
Competitive
  relationships, on bacterial growth,
    172–173
  terminology
    application
      in enzyme studies, 166–171
      in microbiology, 171–173
      in pharmacology, 174–178
      in toxicology, 174–178

  to binding studies, 177–178
  to enzyme inhibition, 169–171,
    181
  to hydrolases, 167–168
  to invertase, 162–165
  to nicotine-curare interaction,
    177–178
  to O_2/CO interactions, 159–162,
    177
  to oxidoreductases, 168–169
  to receptors, 177–178
  to respiration, 159–161, 174, 176,
    177, 181
  to sulphanilamide action,
    171–173, 176–177, 181
  basic concepts, 179
  chemical usage, 165–166
  establishment, 159–161
  first examples, 159–162
  historical development, 159–165
  in drug antagonism, 174–176
  in ecological studies, 179
  in economics, 179–180
Complement proteins, repeating
    homology units, 58–61
Complement system
  activation
    by alternative pathway, 28–29,
      41–43
    by classical pathway, 28–29, 31–41
    of terminal components, 43–47
  activators
    of alternative pathway, 42–43
    of classical pathway, 31–36
  alternative pathway
    components, 30
    inhibition by control proteins, 43
    stabilization by properdin, 43
  anaphylatoxin inactivator, 29, 30
  C1 complex, 31–36, 47–48
    subcomponent C1q B-chain gene,
      cloning of, 52–54
      restriction map, 53
  C3 convertase(s), 29, 38, 39, 40, 41,
    43, 50
  C5 convertase(s), 29, 44, 45, 49, 50
  classical pathway
    activators, 34, 48
    components, 30
  complement receptor 1, 48, 56, 58–61

molecular weights of allotypes, 61
repeating units of, 58–61
structure of, 58–60
components of, 28–30
component C2
  activation of, 39
  genes, 54–55, 56
  homology with unrelated proteins, 59–61
  properties of, 39
  repeating units of, 58–61
component C3
  activation of, 40–41, 43
  cDNA sequence, 61
  covalent binding reaction, 38, 39
  gene, 61
  half life, 40–41
  hydrolysis of, 41
  properties of, 36, 39–40
  reactivity of, 41
  role of, 39–41
  synthesis of, 40
component C4
  activation of, 36–39
  allelic forms, 55–56
  amino acid sequence, 36
  biosynthesis of, 37
  cloning of, 62
  covalent binding reaction, 38, 39
  genes, 54–55
  processing of, 37
  properties of, 36, 39
  synthesis of, 36
component C5
  activation of, 45–46
  amino acid sequence, 45
  cloning of, 62
  properties of, 45
  role of, 45
  sequence of, 62
  synthesis of, 45
component C6
  nature of, 45
  role of, 45–46
component C7, 45
component C8, 45, 46
component C9, 46–47
  cloning of, 62
  in formation of membrane attack complex, 46–47

polymerization of, 46, 49, 62
properties of, 47
control proteins, 30, 41, 42, 43, 47–50
  anaphylatoxin inactivator, 29, 30, 49
  C1-inhibitor, 30, 36, 47–48, 52
  C4-binding protein, 30, 37, 39, 42, 48, 52, 56, 58, 59, 60, 61
    genes, 58
    properties, 48
    repeating units, 58–61
    structure of, 58–60
  evidence for protein super family, 58–61
  factor B, 30, 35, 41, 42, 52
    genes, 54–55, 56–58
    homology with unrelated proteins, 59–61
    repeating units, 58–61
  factor D, 30, 35, 41, 42, 43, 57
  factor H, 29, 30, 41, 42, 43, 50, 52, 56, 59
    gene, 58
    properties of, 48–49
    repeating units, 58–61
    structure of, 58–60
  factor I, 29, 30, 35, 37, 39, 41, 42, 43, 48, 49, 51, 56
  linkage analysis of, 58
  properdin, 29, 30, 43, 49, 50
  S-protein, 29, 30, 44, 47, 49, 52
    cloning of, 62
    model for, 44
    sequence of, 62
decay accelerating factor (DAF), 48, 50–51
genetic defects, 51–52, 54
membrane attack complex (MAC)
  assembly of, 43–46
  control by S-protein, 44, 47, 49, 62
  models for, 44
  molecular weight of, 46
membrane-associated regulatory proteins, 50–51
  complement receptor 1, 48, 50, 56, 58–61
  decay accelerating factor, 50–51
  glycoproteins 45–70 (Gp 45–70), 48, 51

control proteins *continued*
  membrane-bound receptors, 29, 42, 48, 50–51
  membrane lesion formation, 46–47
  molecular biology of, 51–62
  role
    of class III antigens, 55
    of serine proteinases, 35, 36, 48
Corticotropin (ACTH), 71, 76, 87, 91, 92
Corticotropin releasing factor (CRF), 71, 72, 90, 91, 110
Cortisol, in milk protein synthesis, 8, 9
Cytosolic phenolsulphotransferase, 91

D

Debye–Waller factor, 124
Diazepam binding inhibitor, 71
Differentiation antigens, analysis of, 20
Dipeptidyl aminopeptidase IV, 86, 96
Dipole–reorientational relaxations, 122, 146
Dipole–dipole interactions, effect on fluorescence spectrum, 129
Dopamine, 70, 74, 84, 90, 91
Dopamine β-hydroxylase, 88
D-2 dopamine receptor, interaction with metoclopramide, 178
Dynorphins, 71, 76, 77, 79, 87, 98

E

Eledoisin, 79, 80
Endopeptidase—24.11 (EC 3.4.24.11), 97–101, 106–109
  distribution of, 96, 100, 101
  electrophoresis of, 97
  identification of, 100
  immunohistochemical studies, 100
  inhibitor binding, 108
  inhibitors of, 101, 104, 106–109
    acetorphan, 107
    kelatorphan, 107
    leucyl hydroxamate, 108
    natural, 109
    phosphoramidon, 106–108
    thiorphan, 107, 108

isolation of, 97, 100
localization of, 100
monoclonal antibody to, 100
properties of, 96, 104
role of, 107
sites of action, 98–99
specificity of, 100
substrate binding, 108
Endorphin(s), 71, 75, 76, 90, 92, 98, 109
  cross-reactivity with interferon, 109
  synthesis of, 76
Endorphin acetyltransferase, 90
Enkephalin(s), 71, 72, 74, 75, 76, 78, 87, 88, 89, 103, 107, 110
  inactivation of aminopeptidase, 101
  isolation of, 75
  metabolism of, 94–95
  of adrenal medulla, 72, 74, 76, 78, 87, 88, 89
  role of, 76
  source of, 76
  structure of, 75
  synthesis of, 76–79
Enkephalinase *see*
    Endopeptidase—24.11
Enkephalin convertase *see*
    Carboxypeptidase B
Enzyme inhibition,
  allosteric, 171
  analysis of types, 169–171
  development of criteria, 169–171
  graphical methods, 169, 170
  isosteric, 171
  K-system, 171
  V-system, 171
Epidermal growth factor (EGF), 86
  in milk protein synthesis, 8, 9
Ergotamine, 174, 176, 177
Escherichia coli, 168

F

Fibronectin, 48, 62
Fluctuation-dissipation theorem, 146
Fluorescence
  emission, 121
  excitation spectrum of 2,6-TNS, 141–142

in the study of molecular dynamics,
121, 122
intrinsic, 121, 122, 123
lifetime, 131
maxima, dependence on excitation
wavelength, 139–140
polarization, 123, 128–129
probe(s), 121–122, 130
advantages of, 122
aminonaphthalene sulphonates,
122, 138, 139
disadvantages of, 122
cosin, 122, 137
pyrene, 122
mobility,
in phospholipid bilayer, 143
with melittin, 143
2-(p-toluidinyl-naphthalene)-6-
sulphonate (2,6-TNS), 138,
139, 143
with albumin, 141
with apomyoglobin, 139
with β-lactoglobulin, 141
with melittin, 139
quenching, 123, 126–128, 137
mechanism of, 127–128
spectroscopy
edge-excitation method, 134–136
of molecular relaxations, 131–143
Bakhshiev-Mazurenko model,
131–132, 134
general approach, 142–143
in homogeneous broadening
model, 132–136
phase fluorimetric studies, 138
red-edge excitation, 138–142
relaxation shifts, 137
time-resolved, 137–138
FMR Famide, 71, 99

G

GABA see γ-aminobutyrate
GABA transaminase
inhibitors of, 104
mitochondrial, 94
Galactokinase, 170
Galactosyl transferase see
N-acetyllactosamine

synthetase
Galanin, 73
Galleria mellonella, effect of CO, 162,
163
Gastrin, 71, 72, 73, 90, 99
Glucagon, 71
Glucocorticoid receptor, 21
Glutamate decarboxylase, 74
β₂-Glycoprotein I, 58, 59, 60, 61
repeating units in, 59–61
Glycosidases, epididymal, 20
Golgi apparatus, 4, 6, 84, 85, 91
Growth hormone, 71
releasing hormone, 71

H

Haemoglobin, 147, 148, 161, 162, 167,
174
Haloperidol, effect on neuropeptide
production, 91
Heart, as endocrine organ, 83
Heparin, 48
Homogeneous broadening, in
fluorescence spectra, 125
Hormonal control mechanisms, 21
Hormones
circulating, 71
hypothalamic, 71
pituitary, 71
Huntington's Chorea, 110
Hydrocortisone
in milk protein synthesis, 7, 9–10, 11
synergistic effect with prolactin, 9
Hydrolases
competitive relationships, 167–168
inhibition of, 168
Hylambatin, 80

I

Indole, 134, 136
Indoleamines, 95
Induced-fit theory, of enzyme reactions,
170
Inhomogeneous broadening, of
fluorescence spectra, 124,
132–136, 147, 151

Insulin, 7, 8, 71
Interleukin-2 receptor, 58, 59, 60, 61
  repeating units in, 59–61
Invertase (β-D-fructofuranosidase;
    EC 3.2.1.26), 161, 162–165,
    167, 172, 181
  early studies of, 162–165
  inhibition of, 163, 164, 165
    by glucose isomerides, 163, 165
  kinetic studies of, 164

K

Kallikreins, 86
Kassinin, 79, 80, 82
Kramer's equation, 148

L

α-Lactalbumin
  as calcium metalloprotein, 5–7
  as marker of hormone-dependent
      breast cancer, 17
  as model system, 21
  binding to A protein, 6
  Ca^{2+} binding site, 6, 16, 22
  conformational states, 6–7
  covalent structure, 3
  denaturation of, 5–6
  effect on sperm-egg binding, 20–21
  evolution of, 2, 5, 15, 22
  gene(s)
    comparison with lysozyme gene,
        15–16, 21–22
    composition of, 12-15
    conserved regions of, 14, 15
    DNA sequences, 12
    exon–intron organization, 12, 15
    exon structure in relation to
        functional regions, 15–16
    genomic organization, 13
    guinea pig, 12, 13, 14, 15
    human, 12, 13, 14, 15
    isolation of, 12
    rat, 12, 13, 15
    repetitive sequences in, 12, 14
    structure, 11–15
  guinea pig, 2

hormonal control of, 21
  in breast cancer, 16–18
  induction
    by prolactin, 7–8, 9, 21
    in the tammar, 9
  inhibition, by EGF, 8, 9
  in rat epididymal fluid, 19
  interaction with galactosyl
      transferase, 4, 5, 16, 19, 20
  ion binding, 4, 5–7
  MRNA
    in vitro translation, 17, 18
    level
      in breast tumours, 17
      in mammary tissue, 9, 11
    sequences, 12
  physiological function, 2, 4
  production, effect of thyroid
      hormones, 11
  radioimmunological assay, 17
  rat, structure of, 4–5
  sequence-specific cDNA probe, 11,
      12, 17
  similarity to lysozyme, 2, 3, 5, 15,
      21–22
  structure of, 4–5, 22
  synthesis of, 4
    translational control, 9
  X-ray crystallography, 5
α-Lactalbumin-like protein
  function of, 19–21
  identification of, 18–19
  in male reproductive tract, 4, 18–19,
      22
  interaction with galactosyl
      transferase, 19, 22
β-Lactoglobulin, 2
Lactose, synthesis of, 2, 4, 6, 7
Lactose synthase, components of, 2, 5,
    22
α-LAL see α-lactalbumin-like protein
Laminin, 62
Lepidium sativum, effect of CO, 162
[Leu]-enkephalin, 75, 76, 77, 78, 79, 86,
    94, 98
β-Lipotropin, 75
Luteinizing hormone (LH), 71
Luteinizing hormone releasing
    hormone (LH–RH), 71, 90,
    103

Lysozyme, egg white, 2, 3, 5
  covalent structure, 3
  gene
    comparison with α-lactalbumin
      gene, 15–16, 21–22
    exon-intron organization, 12, 15
    exon structure in relation to
      functional regions, 15–16

M

α₂-Macroglobulin, 36, 39, 40, 45, 62
Maltase (α-D-glucosidase;
    EC 3.2.1.20), inhibition of,
    164, 165
Major histocompatibility complex, 52,
    54, 55
  genes,
    class I, 54–55
    class II, 54–55
    class III, 54–55
    genetic map, 55
Melanocyte stimulating hormone
    (MSH), 71, 76, 90, 91, 92
Melittin, 86, 139, 140, 143
  fluorescence maximum, 139, 140
[Met]-enkephalin, 71, 75, 76, 77, 78,
    86, 91, 94, 97, 98
MHC *see* major histocompatibility
    complex
β₂-Microglobulin, 54
Milk proteins
  as model system, 1
  gene expression, 9
  in breast cancer, 16–17
  in guinea pigs, 1–2
  in rabbits, 8–9
  MRNA,
    inhibition of transcription, 9–10
    level in mammary tissue, 9, 11
    stabilization by prolactin, 10
  secretion rates, 9
  synthesis,
    analysis by PAGE 10
    hormonal regulation, 7–11
Molecular relaxations
  fluorescence spectroscopy of,
    131–143
  Bakhshiev-Mazurenko model,

131–132, 134
  general approach, 142–143
  in homogeneous broadening
    model, 132–136
Monoamine oxidase, 94, 95, 104
Morphine, 75, 100
Mössbauer emission, Rayleigh
    scattering of, 147
Motilin, 71
Mouse AtT20 cells, 87, 90
Myasthenia gravis, 55
Myoglobin, 124, 148

N

Naloxone, 107
Natural killer (NK) cells, 109
Nerve growth factor, 86
Neurokinin A (substance K), 71, 79,
    80, 81, 82, 98, 109
Neurokinin B, 71, 80, 82, 98, 101
Neuronal peptides, 73
Neuropeptidases, 94
Neuropeptide(s)
  biosynthesis,
    organization of, 84–86
    signal peptidase, 84
  coexistence with other
    neurotransmitters, 73–74
  concentration in CNS, 73
  distribution of, 71
  functions of, 70–72
  hydrolysis
    by endopeptidase-24.11, 98–99
    by peptidyl dipeptidase A, 102
  identification of, 72–74
  inactivation, 93–104
    by aminopeptidases, 101
    by endopeptidase-24.11, 97–101
    by peptidyl dipeptidase A, 101–103
    mechanism
      at cholinergic synapse, 93, 94
      at GABA synapse, 93, 94
    modes, 94
  in neurological disorders, 110
  localization, 72–74
    by hybridization histochemistry, 73
    by immunohistochemistry, 73
  mammalian brain, 71

metabolism, inhibition of, 104
nature of, 70–72
opioid, 74–79
 effect on immune system, 109
 historical aspects, 74–75
 roles of, 109
precursor processing, 84–93
processing
 by acetylation, 90, 91
 by carboxypeptidase B-like
  activity, 88–89
 by C-terminal amidation, 90
 by methylation, 90
 by phosphorylation, 90
 by tyrosyl O-sulphation, 90
 tissue-specific nature, 91–92
 tryptic-like cleavage, 86–88
  by kallikreins, 86
  by thiol proteinase, 87
proteolytic processing, 84–86
Neuropeptide Y, 71, 73, 74
Neuropeptide YY, 71, 73
Neurotensin, 71, 72, 90, 99, 103
Neurotransmission
 nature of, 70–72
 with classical transmitter, 84, 95
 with peptide transmitter, 84, 85
Nicotinic acetylcholine receptor, 95
Noradrenaline, 70, 74, 84, 85, 88, 109
Nuclear Overhauser enhancement, 125

O

O_2/CO interactions
 in *Galleria mellonella*, 162, 163
 in *Lepidium sativum*, 162
 in man, 160, 161–162
 in the mouse, 161–162
Oestrogen receptors, 17
Opsonization, 43, 51
Oxidoreductases, competitive
  relationships, 168–169
Oxytocin, 71, 72

P

Pancreatic polypeptide (PP), 71
Papain, 146

red-edge effect, 142
Pepsin, 32
Peptidase
 inhibitors,
  design of, 104–109
  uses of, 107, 109
 microvillar,
  antibodies to, 95
  as model system, 95–97
  distribution of, 96
  inhibitors of, 95
  properties of, 95–97
Peptide H1, 73
Peptidergic neurons, 84, 85
Peptidyl dipeptidase A (EC 3.4.15.1),
  101–103, 105–106
 catalytic mechanism, 105
 distribution of, 96, 103
 inhibitors of, 101–103
  captopril, 105, 106, 107
  converstatin, 109
  enalaprilat, 105, 106
  lisinopril, 105–106
 properties of, 96
 role of, 103
 sites of action, 102
Perrin equation, 128
Phosphoglyceromutase, 146
Phosphoramidon
 as endopeptidase-24.11 inhibitor,
  106–108
 as thermolysin inhibitor, 106
 mode of action, 107
Preprotachykinins, 81, 82
Proenkephalin A,
 cDNA, 76
 genes, 76
 proteolytic processing, 86
 sequence, 76
 structure, 76–79
Proenkephalin B,
 genes, 76
 structure, 76–77, 79
Progesterone, in milk protein synthesis,
  7, 9
Proinsulin, proteolytic processing of,
  86, 87
Prolactin, 71
 in milk protein synthesis, 7, 8, 9, 10,
  12, 21

Pronatriodilatin, 83
Pro-opiomelanocortin (POMC), 73, 76, 87, 90, 91
   regulation
      by CRF, 91
      by glucocorticoids, 91
   tissue-specific processing, 91–92
Proteins(s)
   chromophore rotations, 146
   dipoles, 144–146
      effect on enzyme mechanism, 146
      in α-helices, 144–146
      in β-structures, 144–145
      interaction with charges, 145
   dynamics,
      effect on activated processes, 148–150
      limitations of fluorescence studies, 151
   electrostatic interactions, 144, 145
   glycosylation in sperm maturation, 19, 20
   intramolecular dynamics, 126–130
      analysis
         by fluorescence polarization, 128–129
         by fluorescence quenching, 126–128
      reorientation of dipoles, 129–130
   intramolecular motions, 122–126
      analysis with time resolution, 125–126
      by EPR, 126
      by fluorescence depolarization, 126
      by light emission spectroscopy, 126
      by NMR, 125–126
      distribution of dynamic microstates, 122–125
      of dipolar groups, 143–150
      rotations of aromatic groups, 128–129
   microstates,
      distribution of, 122–125
      functional consequences, 146–147
   molecular relaxations
      effect on catalysis, 147–150
      in allostery, 147–150
   synthesis in sperm maturation, 19, 20

temperature-dependent relaxational shifts, 137

R

Red-edge
   effect(s)
      dependence on viscosity, 134, 135
      in multi-tryptophan proteins, 142
      of albumin, 141–142
      origins of, 123, 125, 132–134.
   excitation spectroscopy, 138–142
Relaxational shifts,
   of indole, 130
   of tryptophan, 130
   temperature dependence of, 137
   theory of, 129
Relaxation time(s), 131
   by dielectric dispersion method, 130
   dipolar reorientational
      dependence on temperature,
         for indole, 135, 136
         for tryptophan, 135, 136
Ribonuclease T1, 127, 139
Rimorphin, 79
RNA polymerase III, 14

S

Scatchard analysis, 178
Secretin, 71
Serotonin, 70, 74
δ-Sleep inducing peptide, 71
Somatostatin, 71, 72, 74, 110
Streptobacterium plantarium, effect of sulphonamides, 173
Streptococcus haemolyticus, effect of sulphanilamide, 173
Streptomyces tanashiensis, 106
Substance P, 71, 72, 74, 79, 80, 81, 82, 90, 94, 98, 100, 103, 104, 107, 108, 109, 110
   distribution of, 79
   function of, 79
   historical aspects, 79
   mRNA, 82
   sequence of, 80

Succinate dehydrogenase
  (EC 1.3.99.1), inhibition of,
  169, 172
Sulphanilamide, mode of action,
  171–173
Systemic lupus erythmatosus, 55

                    T

Tachykinins, 79, 80, 82, 90, 98, 100,
  107
  amphibian, 80
  biological activity, 79
  functions of, 79
  interaction with receptors, 82
  mammalian, 80
  molluscan, 80
Thermolysin, 104
Thiorphan, 101, 107, 108
Threonine deaminase, inhibition of, 170
Thrombin, 49
Thrombospondin, 62
Thyroliberin (TRH), 90
Thyrotropin releasing hormone (TRH),
  71, 72, 74
TI-Aj1 see Actinomyces junthinus
  protease inhibitor.
Tri-iodo-thyronine, effect on milk
  protein secretion, 11
t-RNA synthetases, red-edge effect, 142
Trypsin, 86
Trypsin inhibitor, 125
Tryptophan chromophore, 121, 122,
  134, 135, 136, 137, 138, 139,

  140, 141, 144
  dipole–dipole interactions, 121
  effect of red-edge excitation, 134–135
  exciplexes, 121
  fluorescence
    specra, 121
    quenching, 126–127, 151
      by acrylamide, 127, 128, 137
      by ionic quenchers, 127
      by oxygen, 127, 128
    general properties, 121
    intramolecular rotations, 128
Type 1 diabetes, 55

                    U

Uperolein. 80

                    V

Vasoactive intestinal polypeptide
  (VIP), 71, 72, 74
Vasopressin, 71, 72, 83, 109
Viscosity, effect on enzyme reactions,
  147
Vitronectin, 62

                    W

Whey protein, 1–4
  components of, 2
  effect of hormones on induction, 7–11

Cumulative Key Word Index

VOLUMES 1 TO 22

(Volume numbers are shown in **bold** type. Page numbers refer to the first page of the relevant Essay.)

A

ACTIN and MYOSIN, multigene families: expression during formation of skeletal muscle, **20**, 77

ADENOSINE, metabolism and hormonal role, **14**, 82

AFFINITY LABELLING, antibody combining site, **10**, 73

ALDOLASE, structure–function relationships, **8**, 149

ANAPLEROTIC SEQUENCES, role in metabolism, **2**, 1

ANTIBIOTICS, peptides produced by *Bacillus brevis*, **9**, 31

ANTIBODY, topology of combining site, **10**, 73

ARACHIDONIC ACID, alternative pathways of metabolism, **19**, 40

B

BACILLUS BREVIS, peptide antibiotics produced by, **9**, 31

BILIRUBIN, degradation of haem and conjugated, **8**, 107

BIOELECTROCHEMISTRY, **21**, 119

BRAIN, metabolic adaptation in, **7**, 127

BROWN ADIPOSE TISSUE, biochemistry of an inefficient tissue, **20**, 110

C

CARBOHYDRATES, of the mammalian cell surface, **11**, 1

CARBON DIOXIDE, metabolic role in fixation, **1**, 1

CARCINOGENS, CHEMICAL, metabolism in the mammal, **10**, 105

CELL DIFFERENTIATION, nucleic acid synthesis and bearing on, **4**, 25

CELL MEMBRANES, turnover of phospholipids in animal, **2**, 69

CELL SURFACE
human leukaemic cells, **15**, 78
complex carbohydrates of the mammalian, **11**, 1

COLLAGEN (see also Procollagen) structure, **5**, 59

COMPLEMENT, activation and control, **22**, 27

CONTROL, of enzyme synthesis in animal cells, **13**, 39

CYTOCHROME P-450, hepatic, **17**, 85

D

DEOXYRIBONUCLEIC ACID, repair, **13**, 71

DICTYOSTELIUM DISCOIDIUM, study of cellular differentiation, **7**, 87

DIFFERENTIATION
enzymic in mammalian tissues, **7**, 87
polysaccharides and lignin in plant cells, **5**, 89
studies of cellular, **7**, 87

E

ELECTRON TRANSPORT, in *Escherichia coli* mutants, **9**, 1

ELECTRON TRANSPORT CHAIN, photosynthetic, in plants, **1**, 121
ENVIRONMENT, biochemistry of pollution, **11**, 81
ENZYME ACTION, **3**, 25
aldolase, **8**, 149
differentiation in mammalian tissues, **7**, 157
glycolytic-X-ray studies, **11**, 37
regulation of synthesis in animal cells, **13**, 39
study by magnetic resonance spectroscopy, **8**, 79
use of the terms competitive and non-competitive, **22**, 158
ENZYMIC ADAPTATION, in bacteria, **4**, 105
ERYTHROCYTES, mammalian, regulation of glycolysis in, **4**, 69
ESCHERICHIA COLI, a study of electron transport and oxidative phosphorylation, **9**, 1
EVOLUTION
lactose synthetase, **6**, 93
gastric proteinases, **17**, 52
protein, **2**, 147

F

FATTY ACIDS
biosynthesis of unsaturated, **15**, 1
long chain, catabolism in mammalian tissues, **4**, 155
FLUORESCENCE, analysis of protein dynamics, **22**, 120
FUNCTION, gastric proteinases, **17**, 52

G

GAP JUNCTIONS, cell–cell interactions, **21**, 86
GLYCOGEN PHOSPHORYLASE, structure, function and control, **6**, 23
GLYCOLYSIS, regulation in mammalian erythrocytes, **4**, 69
GLYCOLYTIC ENZYMES, X-ray studies, **11**, 37

GLYCOPROTEINS
structure and function, **3**, 105
of mammalian cell surface, **11**, 1

H

HAEM, degradation by mammals, **8**, 107
HEXOKINASES, in animal tissues, **2**, 33
HORMONES, in regulation of glucose metabolism, **3**, 73

I

IMMUNOGLOBULINS, expression of genes, **18**, 1
structure, **3**, 1
INSULIN, proinsulin precursor of, **6**, 69

L

α-LACTALBUMIN, and gene-related proteins, **22**, 1
LACTOSE SYNTHETASE, evolution, structure and control, **6**, 93
LEUKAEMIC CELLS, cell surface, **15**, 78
LIGNIN, metabolism in plant cell differentiation, **5**, 89
LEUKOTRIENES, biosynthesis, **19**, 40
LIPOSOMES, **16**, 49
LIVER
cytochrome P-450, **17**, 85
metabolic control through compartmentation, **7**, 39
biosynthesis of unsaturated fatty acids in mammalian, **15**, 1
LYSOSOMES, **12**, 1

M

MAGNETIC RESONANCE SPECTROSCOPY, application to the study of enzymes, **8**, 79
MECHANISMS OF ACTION, gastric proteinases, **17**, 52
MEMBRANE FLUIDITY, structure and dynamics of membrane lipids, **19**, 1

MEMBRANES, of bacteria, **8**, 35
MEMBRANE STRUCTURE, spectroscopic studies, **11**, 139
MESSENGER RNA, mammalian, **9**, 59
METABOLISM
  action of hormones in glucose metabolism, **3**, 73
  adaptation in the brain, **7**, 127
  control of, by adenosine, **14**, 82
  of animal phospholipids, **2**, 69
  catabolism of long chain fatty acids, **4**, 155
  control in liver through compartmentation, **7**, 39
  in lymphocytes, **21**, 1
  of pyruvate in mammalian tissues, **15**, 37
  role of CO_2 fixation in, **1**, 1
  role of anaplerotic sequences in, **2**, 1
  of starch, **10**, 37
MICROTUBULES, and intermediate filaments, **12**, 115
MICROVILLI, enzymology and molecular organization, **14**, 1
MITOCHONDRIA, intracellular membrane function, **6**, 1
MUCOUS GLYCOPROTEINS, **20**, 40
MUSCLE, mechanism of contraction, **1**, 29; **10**, 1
MUTAGENS, CHEMICAL, metabolism in the mammal, **10**, 105

N

NEUROPEPTIDES, processing and metabolism, **22**, 69
NEURON, the neurosecretory, **14**, 45
NUCLEAR MAGNETIC RESONANCE, uses in biochemistry, **19**, 142
NUCLEIC ACIDS, sequence determination, **1**, 57
  synthesis in embryos, **4**, 25

O

OESTROGENS, mechanism of action, **20**, 1
ORAL CONTRACEPTIVES, steroidal agents, **2**, 117
OXIDATIVE PHOSPHORYLATION, **1**, 91
  in *Escherichia coli* mutants, **9**, 1

OXYGEN, activation, **17**, 1
  transport and storage proteins of invertebrates, **16**, 1

P

PASTEUR EFFECT, respiration and fermentation relationship, **8**, 1
PHOSPHOLIPIDS, metabolism in the animal, **2**, 69
PHOTOSYNTHESIS, electron transport chain, **1**, 121
PLANTS, polysaccharide and lignin metabolism in cell differentiation, **5**, 89
  α-amylase and the hydrolases in germinating cereal seeds, **18**, 40
POLYSACCHARIDES, metabolism in plant cell differentiation, **5**, 89
POLYMERIZATION, relation between direction and mechanism, **4**, 1
PROINSULIN, precursor of insulin, **6**, 69
PROCHIRALITY, in biochemistry, **9**, 103
PROCOLLAGEN, biosynthesis, **12**, 77
PROSTAGLANDINS, biosynthesis, **19**, 40
PROTEINASES, structure, function, evolution and mechanism of action of gastric, **17**, 52
PROTEIN, degradation of intracellular, **13**, 1
PROTEINS
  biosynthesis of multi-chain, **5**, 139
  dynamics of, **22**, 120
  mechanisms of evolution, **2**, 147
  oxygen storage of invertebrates, **16**, 1
  translational regulation in eukaryotes, **18**, 79

R

RIBOSOME, evolutionary diversity, **21**, 45

S

SECRETION, the neurosecretory neurone, **14**, 45

SKIN, biochemistry of human, **7**, 1
SPECTROSCOPY, studies of membrane structure, **11**, 139
STARCH, structure and metabolism, **10**, 37
STEROID HORMONES
  actions of, **5**, 1
  oral contraceptive agents, **2**, 117
  intracellular reception, **12**, 41
  production by dispersed cells, **16**, 99
STRUCTURE
  of gastric proteinases, **17**, 52
  of glycoproteins, **3**, 105

of immunoglobulins, **3**, 1
of mucous glycoproteins, **20**, 40
of starch, **10**, 37

T

TEICHOIC ACIDS, bacterial cell walls and membranes, **8**, 35
TETRAPYROLLES, and oxygen activation, **17**, 1
THROMBOXANE, biosynthesis, **19**, 40